T0292878

Toward Human-Level Artificial Intelligence

Is a computer simulation of a brain sufficient to make it intelligent? Do you need consciousness to have intelligence? Do you need to be alive to have consciousness? This book has a dual purpose. First, it provides a multi-disciplinary research survey across all branches of neuroscience and AI research that relate to this book's mission of bringing AI research closer to building a human-level AI (HLAI) system. It provides an encapsulation of key ideas and concepts, and provides all the references for the reader to delve deeper; much of the survey coverage is of recent pioneering research. Second, the final part of this book brings together key concepts from the survey and makes suggestions for building HLAI. This book provides accessible explanations of numerous key concepts from neuroscience and artificial intelligence research, including:

- The focus on visual processing and thinking and the possible role of brain lateralization toward visual thinking and intelligence.
- Diffuse decision making by ensembles of neurons.
- The inside-out model to give HLAI an inner "life" and the possible role for cognitive architecture implementing the scientific method through the plan-do-check-act cycle within that model (learning to learn).
- A neuromodulation feature such as a machine equivalent of dopamine that reinforces learning.
- The embodied HLAI machine, a neurorobot, that interacts with the physical world as it learns.

This book concludes by explaining the hypothesis that computer simulation is sufficient to take AI research further toward HLAI and that the scientific method is our means to enable that progress. This book will be of great interest to a broad audience, particularly neuroscientists and AI researchers, investors in AI projects, and lay readers looking for an accessible introduction to the intersection of neuroscience and artificial intelligence.

Eitan Michael Azoff, PhD, is Chief Analyst at Cloud and Data Center Research Practice, Omdia, part of Informa.

Toward Human-Level Artificial Intelligence

Artificial Intelligence

How Neuroscience Can Inform the Pursuit of Artificial General Intelligence or General AI

Eitan Michael Azoff

CRC Press
Taylor & Francis Group
Boca Raton London New York

CRC Press is an imprint of the
Taylor & Francis Group, an **informa** business

Designed cover image: Eitan Michael Azoff

First edition published 2025
by CRC Press
2385 NW Executive Center Drive, Suite 320, Boca Raton FL 33431

and by CRC Press
4 Park Square, Milton Park, Abingdon, Oxon, OX14 4RN

CRC Press is an imprint of Taylor & Francis Group, LLC

ISBN: 9781032831305 (hbk)
ISBN: 9781032829074 (pbk)
ISBN: 9781003507864 (ebk)

DOI: 10.1201/9781003507864

Typeset in Sabon
by codeMantra

For Josefina

Contents

Preface   *xi*

Acknowledgments   *xv*

PART ZERO
AI level setting   I
Introduction 1

1 AI and machine learning   2

2 Elements of neural networks history   5

    2.1   *Introduction 5*
    2.2   *From birth of neural networks to an AI winter 6*
    2.3   *Backpropagation 7*
    2.4   *Deep learning 8*

PART ONE
Neuroscience implications for HLAI   II
Introduction 11

3 Brain properties   12

    3.1   *Introduction 12*
    3.2   *The neuron 12*
    3.3   *Action potential 16*
    3.4   *Dendrites 19*
    3.5   *Glial cells 20*
    3.6   *Grid neurons 22*
    3.7   *Mirror neurons 23*

  3.8   *Electrical communication in the brain:*
        axons, dendrites, nucleus, and synapses 23
  3.9   *Molecular communication in the brain:*
        neuromodulators and neurotransmitters 27
  3.10 *Human memory 29*
  3.11 *Brain lateralization 33*
  3.12 *Brain folds and neocortex columnar structure 35*
  3.13 *Early brain development 36*
  3.14 *Brain activity, sparsity, and normalization 39*
  3.15 *Neuron mixed selectivity 40*
  3.16 *Neural oscillations 41*

4 Cognitive processes   45

  4.1   *Introduction 45*
  4.2   *Cognition 45*
  4.3   *Human consciousness 46*
  4.4   *Animal consciousness 49*
  4.5   *Sensory thinking 51*
  4.6   *Vision and the rules of perception: visual intelligence 52*

5 Time and space in the brain   55

PART TWO
Theories, models, and algorithms   59
  Introduction 59

6 Theories of consciousness   60

7 Neurorobotics: embodied AI   62

8 Engineered brain architectures   65

  8.1   *Introduction 65*
  8.2   *Cognitive architectures 65*
  8.3   *Adaptive resonance theory model of the brain 72*
  8.4   *Harmonic oscillator recurrent neural networks 76*
  8.5   *Numenta AI models 79*
  8.6   *Deep learning neural networks 83*
  8.7   *Biologically plausible models 93*
  8.8   *Hyperdimensional computing 103*

9 AI hardware   105

   9.1   Introduction 105
   9.2   Neuromorphic processors 105
   9.3   NeuRRAM analog chip 110
   9.4   Nvidia AI GPUs 110
   9.5   In vitro neurons learn to play Pong 111

PART THREE
Speculations toward human-level AI   **115**
   Introduction 115

10 The possibility of creating HLAI   117

   10.1   Three types of HLAI 117
   10.2   Neuroscience inspiration 119
   10.3   Consciousness and HLAI 121
   10.4   Visual thinking and consciousness 122
   10.5   Brain lateralization, visual processing, and intelligence 126
   10.6   Memory in HLAI systems 127
   10.7   The executive seat in the brain versus
          diffuse decision making 128

11 Methods to build HLAI   130

   11.1   Introduction 130
   11.2   The intelligent robot as scientist 132
   11.3   Evolving intelligent systems 136
   11.4   LLM intelligence and potential for HLAI 137
   11.5   The key attributes and tests of an A/HLAI system 138

12 Beyond HLAI   144

   Epilogue   146
   Appendix: notes on the scientific method   147
   Glossary   149
   References   150
   Author: Eitan Michael Azoff   169
   Index   171

Preface

The foremost assumption made in this book is that it is possible for us to understand how the human brain works if we crack the neural code, i.e., how the human brain encodes the sensory information it receives, and moves information in the brain to perform cognitive tasks, such as thinking, learning, problem solving, internal visualization, and internal dialogue. A second key assumption is that in breaking through with such understanding we will then have the knowledge to create a human-level artificial intelligence (HLAI): a machine that can perform cognitive tasks at the level that the best humans can achieve. There's a lot wrapped up in these statements, for example:

1 Is computer simulation of a brain sufficient to make it intelligent?
2 Do you need consciousness to have intelligence?
3 Do you need to be alive to have consciousness?

If your answer to the first of these questions is no, then I hope this book will open your views to counter arguments; however, if your answer is yes, then I trust this book will provide material for your research or feed your curiosity. I will return to these questions in Part Three, and the question of whether a computer simulation can possess HLAI.

To speculate, an HLAI machine will be able to open new dimensions of capabilities. Our first tentative steps toward HLAI may well be primitive, possibly as slow (compared to modern computers), or slower than the human brain or brains of simpler animals with fewer neurons than humans. But once we crack the neural code we will engineer faster and superior brains with greater capacity, speed, and supporting technology that will surpass the human brain. This is even before transcendency or the so-called singularity (Kurzweil, 2005) is reached, where the AI machines create the next generation of intelligent machines that reach even higher levels of intelligence.

I have opted not to discuss whether these prospective developments in AI are good for humanity or not except to say that I think such developments in HLAI can offer huge benefits for mankind. However, society must also act to control this technology and prevent its misuse. Knowledge is value free; it is up to society to ensure it is used morally and ethically. These are deep and complex topics that are beyond the scope of this work.

There is a general principle assumed here that all life forms on Earth share common features in their makeup, that evolution reuses what it has created, though typically incrementally in more evolved forms. This assumption is borne out in neuroscientific research. For example, it's what gave Eric Kandel the Nobel prize for his work on memory, which was conducted on a sea snail, but the principles learned in that research are relevant to the human brain. Hence research on simpler life forms to understand how their brain works is a useful approach to understand the workings of the human brain. Take the honeybee, with some 1 million neurons, it is capable of a surprising degree of higher cognitive functions, including the ability to cope with the concept of "sameness", recognize human faces, use top-down visual processing, as well as solve complex maze-type problems and show context-dependent learning (Rogers et al., 2013a).

There is a huge amount of scientific research on the brain conducted in many disciplines: in the many branches of neuroscience, psychology, and AI research, but the scientists and engineers work largely within their silos, writing in their house journals, and attending their disciplines' favored conferences. There isn't enough cross-over. It is a kind of scientific tower of Babel where scientists rarely come up to the surface to communicate with scientists from other disciplines. This will have to change because I'm convinced that only a multi-disciplinary approach will crack the neural code.

While some AI researchers have the goal of achieving HLAI, neuroscientists working on understanding the human brain have more varied aims, and some can be dismissive of such AI research because of its primitiveness compared with the sheer complexity of biology, and even the simplest cell life form. But many neuroscientists (including neuropharmacologists, neurophysiologists, neuroanatomists) are not trained in computation and may not appreciate clues their work can offer as useful to AI researchers. Of course, there are disciplines such as computational neuroscience (computational neurophysiology, evolutionary cognitive neuroscience, computational cognitive science) that may speak the language of AI researchers (i.e., mathematics and simulation) but have radically different research aims. And then cognitive psychologists form another research group altogether. The proliferation of specialisms is multiple and creates barriers for useful information to flow between them.

There is another challenge, nicely put by Terrence Sejnowski (Anderson & Rosenfeld, 2000) who moved across disciplines, from relativistic physics to neuroscience and biology to neural modeling, that once you start experimenting in biology the complexity and detail is so deep that you don't get a chance to rise above it to find system level answers. The challenge becomes understanding what detail is important to carry forward into an engineering model. A multi-disciplinary approach is essential, and the role of AI modeling is to tie the necessary details together into a simulation of the brain that emulates intelligence.

This is the motivation for this book: to bring together the clues to be found across all the science concerned with the brain, gather them in the whole to inspire the next steps in AI research. This book does not crack the neural code, but I would like to inspire the next generation of scientists and engineers to (a) make them aware of the challenge, (b) provide a basis for how this challenge can be solved, and (c) collect the relevant clues scattered across multiple papers and books and reduce the barriers of the scientific tower of Babel. I believe that we are at a critical point in the history of AI, where the AI community is growing at a fast pace, largely preoccupied with narrow AI applications, while neuroscientific data is growing at a rapid pace thanks to ever evolving non-destructive measuring technology (see Appendix). The research cited in this book from across multiple disciplines should provide a convenient starting point for AI researchers building HLAI.

The readership this book aims for runs across anyone interested in the topic of where AI goes next: AI researchers, neuroscientists, students, research budget holders, venture capitalists, and the interested general reader. This book is divided into four parts:

- Part Zero provides a level setting for definitions of what I mean by AI, machine learning, deep learning, etc. It provides a little history as well to put our current state of knowledge in AI into perspective so that the challenges ahead can be put into historical context.
- Part One pulls out of the neuroscience research literature useful pieces of information that should inform future AI models. It is all about facts, and is evidence based.
- Part Two covers the theories (perhaps more accurately called hypotheses) being pursued by first neuroscientists and then AI researchers on how the brain works.
- Part Three is speculative in nature, pulling bits from earlier chapters and trying to make sense of them holistically, in a logical manner, to guide research toward HLAI. For those AI researchers building HLAI systems I offer a series of test questions to compare with your model. This test is designed to follow more closely our knowledge of the human brain and what is likely to be needed in an HLAI model.

Finally, note there may be relevant research that I have missed or statements that later prove incorrect. Do write to me at my email below if I have missed research work or made incorrect statements. Some material may become dated and prove incorrect over time, but that is normal, it is how science progresses. My aim is to draw a line in the sand and say: this is where we are at the time of writing, as we make progress in our understanding, we will redraw the line closer to our goal.

Note, there are essentially three purposes to the use of references here:

- To describe, or quote directly from a relevant published work.
- To refer the reader to a paper, typically a review, or a book, that can help further explain the topic under discussion.
- In some cases, to attribute historical priority. However, this is not a work of history and is not meant to be comprehensive or exhaustive in respect of prior art.

Finally, please leave a review where possible and tell me what you liked or did not like in this book.

Dr. E. Michael Azoff
Newark, UK
Email: ema@hmnlvl.ai
Book web site: www.hmnlvl.ai

Acknowledgments

I work outside academia, and it would not be possible to have had access to all the neuroscience and AI research papers without the policy of online open access that so many authors subscribe to (for a few papers that were not easily accessible, their authors have been kind enough to send me copies). This book is a direct product of open access and I owe great thanks to its existence. Thanks also to Stephen Grossberg for a detailed review of my coverage of his research. I'm grateful to Frank Ritter for a review of the manuscript and many helpful comments. I would like to thank the authors who allowed their images and figures to be reproduced. A big thank you to Elliott Morsia my editor at Taylor & Francis for supporting publication of this work.

Part Zero

AI level setting

INTRODUCTION

One important confession I wish to make at the outset: I use the term human-level AI (HLAI) in respect of machine intelligence that matches human intelligence – the same level as a human. But you won't find a definition here of what I mean by human-level intelligence. As it is (mostly) humans reading this book, I will assume we all know what we mean by *our* intelligence then I can perhaps be forgiven the lack of definition. The same applies to the other widely used term, artificial general intelligence (AGI) or general AI – it suffers from the same lack of hard definition. And the reason there is no hard definition is first, that how intelligence is created, cracking the neural code, understanding how the brain works – the mission of this book – is currently unknown. And second, human intelligence has multiple dimensions (computational, mathematical, logical, emotional, social, artistic, and more) and this is an aspect beyond the scope of this book. Note: it will become clear in Part Three why I have avoided AGI in favor of HLAI as it allows useful labeling of variations. For measuring intelligence, I refer the reader to Jose Hernandez-Orallo's comprehensive work (2017).

This book is aimed at readers who are currently up to date with AI today, as the contents are focused on next steps for AI and the AI of the future. However, for the sake of completeness I provide what I call level setting. The idea is to explain what I mean by AI so that the reader can relate the discussion here with reference to their own interpretation of AI. The chapter on neural networks history is brief but I trust useful to the reader, as I've seen confused references to concepts such as AI winter and the origin controversy of backpropagation, a crucial algorithm for training neural networks.

DOI: 10.1201/9781003507864-1

Chapter 1

AI and machine learning

Stuart Russell and Peter Norvig (2020) in their book *Artificial Intelligence: A Modern Approach* reviewed the literature to see how AI researchers defined their field and came up with essentially the scheme in Figure 1.1. The two extremes are the lower left, a machine that performs like a human, and the upper right, a machine that can think rationally. These dimensions of AI go beyond thinking and into the realm of a robot that can act, move, and perform physically like a human across all senses: vision, sound, smell, motion, and tactile dexterity. There are AI researchers who argue that as intelligence evolved from our physical interaction with the world, we should investigate AI that is similarly connected with the physical world. What is clear from this analysis is that to be human is to include our emotions, but we may wish to build AI that acts purely rationally without the interference of emotions; hence, the x-axis in the figure separates out human-like and purely rational behavior.

The aim for many AI researchers is to build intelligent machines that achieve parity with human intelligence and then go beyond to higher levels of intelligence, which the machines may achieve but is beyond human capacity (the term transcendence or singularity is used). The timeline depicted in Figure 1.2 charts our progress toward HLAI and our current state of progress, which I have positioned at just beyond narrow AI. There is no standard definition of narrow AI, but essentially, the AI application does one task well but does nothing else (hence "narrow"). Following the real-world successes of deep learning, generative AI, and its applications with large language models that can perform well on multiple benchmarks including logical reasoning (see Section 8.6.2), I have marked the current state of the art as a little beyond narrow AI.

There are many algorithms being developed by AI researchers, some going back to the invention of backpropagation and still in use today, to new research continually being published. Machine learning (ML) is a sub-branch of AI, and has transitioned from research into real-world applications. Figure 1.3 shows the subject hierarchy for connectionist models,

DOI: 10.1201/9781003507864-2

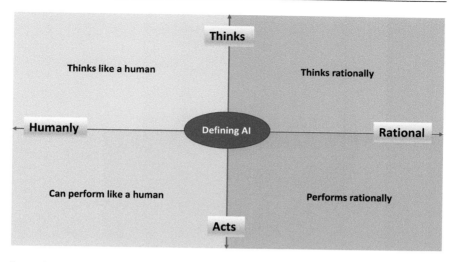

Figure 1.1 How AI researchers define AI.
Source: E. M. Azoff.

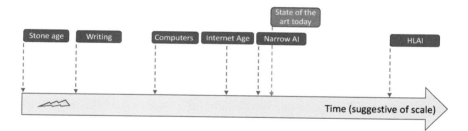

Figure 1.2 The current state-of-the-art in AI.
Source: E. M. Azoff.

focusing on ML. ML is a broad set of algorithms and techniques, within which brain-inspired research is the oldest and original segment and neural networks is one of the most prominent technologies. Within neural networks research deep learning (DL) is the most recent and successful. Other notable technologies (not shown) are not-connectionist and include symbolic and expert systems, Bayesian probabilistic learning systems, natural language processing, and more. A graphic depicting all the research areas would include multiple overlaps as hybrid systems are also possible.

The latest research in generative AI falls into the DL category and is covered in Section 8.6.

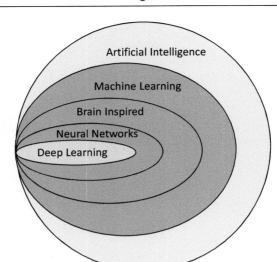

Figure 1.3 Deep learning and how connectionist research topics in AI relate to each other. Other non-connectionist areas of AI research are omitted.

Source: E. M. Azoff.

Chapter 2

Elements of neural networks history

2.1 INTRODUCTION

The following high-level and brief history of the development of neural networks touches on two of its controversies, the cause of the first, so-called, AI winter, and the origins of backpropagation, a method for training neural networks that is still important today for deep learning. These are controversies at least to the participants in the history, but it is important to know some of this background to get an understanding of how the art and science of neural networks evolved to where it is today, with the current success (at least in practical engineering terms) of deep learning neural networks (DLNNs) and generative AI.

Readers interested in this history should read the original papers reproduced in Anderson and Rosenfeld's *Neurocomputing* (1989), and their *Talking Nets* (2000) book subtitled *An oral history of neural networks* where they interview the key figures. Apart from supervised neural network training algorithms, such as backpropagation, there are other training methods, such as unsupervised and reinforcement, as well as alternate architectures, such as recurrent networks and Bayesian belief networks. Grossberg's Adaptive Resonance Theory is discussed in Section 8.3. For a complete overview of the techniques and models developed in the period discussed below, see books by Hagan et al. (2014), Haykin (2009), Hertz et al. (1991), and Levine (2019).

Finally, to mention this is a rather North American-centric history, and for an alternative account, the reader should consult Jurgen Schmidhuber (2015a, 2015b, 2022). In support of this account of neural network history in the period covered, AI research in North America had the greatest influence on the field, as well as its application outside academia, not least due to the multiple centers of research and availability of research grants to push the research forward.

DOI: 10.1201/9781003507864-3

2.2 FROM BIRTH OF NEURAL NETWORKS TO AN AI WINTER

Modern AI history starts with the work of McCulloch and Pitts (1943) where the brain is modeled with an artificial (vastly simplified) neural network that computes with logical calculus. This was a failed model but one that inspired Frank Rosenblatt (1962) to create the feedforward neural network, named perceptron. Rosenblatt's experiments/simulations of the perceptron were limited to two layers, input and output, because he lacked a method to train any middle or hidden layers, but he understood the advantages of a multi-layer perceptron and the concept of taking the error at the output (i.e., comparing the output with a known target) and somehow feeding this back to improve the model.

Interestingly, Marvin Minsky who came to represent the symbolic branch of AI research had started out working with the perceptron in his early research in the late 1950s. By the 1960s, Minsky had moved on to symbolic AI, where machines work with high-level symbolic (human readable) forms of logic. In contrast, the perceptron creates a distributed representation of information across the neurons, and this form of connectionist logic acts as a black box: opaque to direct human interpretation.

Minsky and colleague Seymour Papert became quite hostile to connectionist AI and did their best to discourage research in this field. In Jack Cowan's recollections (Anderson & Rosenfeld, 2000), Minsky and Papert were disseminating research preprints to the AI community and talking about the limitations of the two-layer perceptron in the years leading up to publication of their book *Perceptrons* (Minsky & Papert, 1969). According to Robert Hecht-Nielson (Anderson & Rosenfeld, 2000), Minsky and Papert's colleagues advised them to tone down the vitriol in the book to the point that it was dedicated to Frank Rosenblatt!

Minsky and Papert's book explores the limitations of the two-layer perceptron, and the outstanding problem of a training rule for hidden layers. Bernard Widrow in his recollections (Anderson & Rosenfeld, 2000) notes that by the time *Perceptrons* was published, there was little work going on with neural networks. The connectionist research community faced an "AI winter", with funding severely limited for any proposal mentioning neural networks.

Two pockets of research activity that continued to work with neural networks were centered around Stephen Grossberg, and the cognitive psychology group centered around David Rumelhart. Rumelhart had organized a group reading of the *Perceptron* book soon after publication which ignited his interest in neural networks. He chose the name for his research, parallel distributed processing (PDP), and with colleagues James McClelland and the PDP Research Group they published three volumes: PDP volumes 1 and 2 (Rumelhart et al., 1986b), and *Explorations in PDP: A handbook of models, programs, and exercises* (McClelland & Rumelhart, 1988). The release of neural network code in volume 3 was an early instance of sharing code, available in the book on a floppy disc (this was of course the personal computer era, pre-internet, and no GitHub). Sharing of code is a common feature in the AI community today, and I believe a key reason for the rapid evolution of the connectionist field, and an early example of open source software. Rumelhart was a strong advocate of openness and sharing of his research (Anderson & Rosenfeld, 2000).

2.3 BACKPROPAGATION

A key result published in PDP volume 1 is the chapter by David Rumelhart, Geoffrey Hinton, and Ronald Williams: "Learning internal representations by Error Backpropagation", also published as a paper in Nature (Rumelhart et al., 1986a): this was the backpropagation algorithm for training the internal "hidden" layers of a multi-layer feedforward neural network (i.e., perceptron). The backpropagation idea belonged to Rumelhart. This result gave birth to the modern era of neural networks, because multi-layer neural networks had a lot of power and capability.

Minsky and Papert's negative thinking of the perceptron no doubt led to their missing the essential switch (or cognitive leap) that makes backpropagation possible. Researchers in the field were using threshold logic for the neuron's function. The step function threshold has infinite gradient and so can't be differentiated. The simple expedient of changing the step function into an approximate one but which was smooth and differentiable, such as the sigmoid function, meant that it was possible to apply the chain rule of calculus to backpropagate the error function at the output back through hidden layers.

This leads to the second so-called controversy in the early days of AI: who discovered/invented backpropagation? Rumelhart (Anderson & Rosenfeld, 2000) makes two points worth noting: first, the simplicity of the threshold to sigmoid switch may explain why backpropagation was independently discovered multiple times, and second, many of the earlier discoveries did not go any further for lack of implementations, whereas the work of Rumelhart and team was coded into computer simulation and the code was disseminated.

In terms of precedence, Schmidhuber (2015a, b) has unearthed backpropagation-like origination papers that go back to the 1960s, e.g., Ivakhnenko and Lapa (1965). Shunichi Amari (1967) independently published a paper with the concept. Paul Werbos's (1974) PhD thesis independently rediscovered backpropagation, as did Yann Le Cun (1985). With some irony, in 1971–1972, Werbos was scouting for a patron so he could write up his PhD thesis and approached Minsky, who Werbos says refused to have anything to do with a method that involved a differentiable neuron function (Anderson & Rosenfeld, 2000).

Minsky and Papert published a second edition of their Perceptron book (Minsky & Papert, 1988) answering back to Rumelhart and team in the epilogue that many of the theoretical issues addressed in their book were still relevant, while acknowledging their failure to anticipate backpropagation, but by this stage, the AI research "primacy" baton had quite definitely passed to the connectionist camp: there was a huge burst of activity and important research funding bodies such as Defense Advanced Research Projects Agency (DARPA) were championing this branch of AI once more.

One final note concerning backpropagation prior art: both Stephen Grossberg and Paul Werbos faced major obstacles to getting their work published in major journals because of resistance or misunderstanding of their research work. Grossberg was years ahead of his time and Werbos persevered and ingeniously applied his algorithms to economic and political applications.

By the mid-1990s, the optimism in the connectionist camp faded on the back of quite modest success with the neural networks of the day, measured by practical, real-world applications as well as progress in pure AI research. That said, there were spikes of application success, for example, the most successful hedge fund in history, Jim Simon's Renaissance Technologies was launched in 1988 and used pure quant methods including the neural networks of the day (Zuckerman, 2019). Robert Hecht-Nielsen (Anderson & Rosenfeld, 2000) remembers there *were* commercial successes, but these were not generally known, one example was automated optical character recognition that was used by many businesses. With another AI winter settling over the connectionist community and funding drying up, a small band of researchers, as before, continued to research with the ruse of label change; this time, it was called deep learning (Le Cun et al., 2015; Schmidhuber, 2022).

2.4 DEEP LEARNING

The research laboratories run by Geoffrey Hinton at the University of Toronto and Yoshua Bengio at the Université de Montréal made significant breakthroughs in the period 2010–2012, a key one was porting their DLNN algorithms on to a new generation of graphical processing units (GPUs) that enabled general programming. This chip was Nvidia's general-purpose graphics processing unit (GPGPU), which led to a "Cambrian explosion" of AI accelerator processors in the market, in many forms and architectures. For anyone not familiar with these chips, the massively parallel processing cores in a GPU, ordinarily used for video pixel processing, are harnessed for calculating matrix multiplication and accumulate operations needed in feedforward and backpropagation cycles when training neural networks – for AI applications, the label "graphics" is redundant. These hardware accelerators reduced training in very deep network models from months and weeks to days and hours and less (depending of course on the size of the model).

Working with neural networks is as much an art as science. How you architect the model – the number of layers, the number of neurons per layer, the connection topography, the functions used in the neurons, etc. was based on heuristic rules gained through mostly trial and error simulation experiments (empirical research, to use another term). Deep learning modeling is similarly based on heuristic rules, but many of these overturn the previous generation rules. For example, in the 1990s, neural networks were typically built as small as possible to fit the requirements and overcame the "curse of dimensionality": if the number of model free parameters were equal to or greater than the number of available data points for training, then the model did not learn to generalize, it just imprinted the unique set of training data points and then performed poorly on new unseen data. With the availability of big data that problem disappears, but even where data is limited modern DLNNs are big but use techniques like pruning and sparse connectivity to build very deep layered models that don't fall under the curse of dimensionality.

The AI research community continues to be split between those pursuing symbolic approaches and those pursuing artificial neural networks. Symbolic AI ideas continue in cognitive architecture research, covered in Section 8.2. The work of the symbolic AI community also led to expert knowledge systems and commercial business rules engines that are in use today. A major benefit of symbolic systems is their transparency, the rules make clear how a result is achieved. Another legacy is probabilistic programming, which combines techniques from statistics and computer science for inference applications (inference mode is where a trained AI model is used in production). DLNNs first made an impact in 2011 when they were used in an image classification competition and beat all the expert algorithms that researchers had developed over decades. To the public the most spectacular consequence of deep neural networks is its foundation for generative AI and large language models (LLMs).

A major difference between the historic years and the present is the embracing of deep neural networks and machine learning by industries and enterprises, which has led to huge amounts of venture capital pouring into the field. The old infighting between academic AI communities for funding pales compared to the VC funding available today for AI technology startups.

The history of DLNNs and the birth of the modern era of generative AI is picked up again in Section 8.6.

Part One

Neuroscience implications for HLAI

INTRODUCTION

In this part, neuroscience research is highlighted that casts light on any aspect of the brain that helps understand how the brain thinks. Readers who need a broad introduction to neuroscience should consult standard texts (Martin et al., 2021; Levitan and Kaczmarek, 2015; Kandel et al., 2013), but note some of the groundbreaking research quoted here is too recent to appear in textbooks – neuroscience research has expanded considerably, and the research output is fast paced. Another basis for this selection of material is to undo ideas about the brain that have caught the public imagination, but recent research shows it to be incorrect. One example is Paul MacLean's triune brain hypothesis, conjectured in the 1960s, which sees the brain as a successive, *serial* evolution of three main feature sets, building on top of each earlier version. According to this theory, first the reptilian brain evolved, largely comprising instinctual, survival behaviors, then came the paleomammalian brain (limbic system) that introduced emotions and greater memory capacity, and finally neo-mammalian, which gave higher mammals the neocortex and rational thought. However, today this model is considered inaccurate, brain evolution throughout the animal kingdom and not just mammals, is more complex. Research shows the brain evolved all three triune properties *concurrently* in ever greater degrees of sophistication over time in higher animals and most evolved in mammals (Striedter and Nothcutt, 2020).

The brain's language capability is largely underplayed in this survey for the reason explained in Part Three.

DOI: 10.1201/9781003507864-4

Chapter 3

Brain properties

3.1 INTRODUCTION

Neurons are found throughout the central nervous system. Those found in the spinal cord fall into three types (Moini and Piran, 2020):

- *Sensory*: sensory neurons are the first nerve cells that respond to sensory information from the environment, relaying the information to the rest of the nervous system.
- *Motor*: these neurons are found in muscles, glands, and organs, taking inputs from the spinal cord and controlling the muscles. There are also motor neurons in the brain that connect to the spinal cord.
- *Interneuron*: these neurons, which can number 100 billion, connect the sensory and motor neurons. These numerous neurons can form complex circuits between themselves.

The types of neurons in the brain run to hundreds and there is no established classification system yet devised. They also vary in which neurotransmitters they respond to. Moreover, neurons in the brain sit within a zoo of other cell types, e.g., macrophages and microglial cells, in a sea of gray or white matter crisscrossed by blood vessels. These different cells play vital roles. For example, oligodendrocyte cells live in this sea of matter and fluids and produce the myelin sheaths that grow around neuron axons, acting as electrical insulators that improve the transmission speed and signal strength of the action potentials (or electrical spikes, we will use the label interchangeably) that travel down the axons. Glial cells such as astrocytes connect to blood vessels and interact with neurons in synapses.

3.2 THE NEURON

The neuron is a cell with a nucleus (eukaryotic cell) in contrast with single-cell life forms such as bacteria that are smaller (by a factor of 100 to

                                    DOI: 10.1201/9781003507864-5

10,000) and have no nucleus (prokaryotic cell). The simplest living creatures on our planet are single-cell life forms, and there is a direct line of evolution from these cells to multi-cellular life forms, including humans. Humans can be thought of as cellular eco-systems, the body comprises some 30 trillion cells, and there are some 30 trillion single-cell "foreign" microbes in our gut, without which we could not exist. Neuroscience experiments on the simplest cells to higher animals are relevant for humans because all life forms on our plant share the same chemistry, so that experiments on the synapse of a sea snail are relevant to the human brain synapse (Kandel, 2006). To take another example, in the 104 amino acid sequences of the Cytochrome c protein, 26 are the same in all animal species, from bacteria to human, retaining it across 1.5 billion years of evolution.

Dennis Bray (2009) points out that single-cell life forms make sophisticated decisions through molecular computation. The role of DNA in genes is to create proteins, and these proteins are akin to computer algorithms. A protein can exist in two states: a stable low energy state and a transient high energy state. Flipping between these two states acts as a switch: IF <in state-1 and attached to molecule B> THEN <switch to state-2 and act as functioning enzyme converting molecule A to A'>. This topic of protein switching is known as allostery (Liu, 2016). In a simple bacterial cell, the number of possible states that all the proteins can take is virtually infinite (Bray, 2009).

The "neuron doctrine" in the history of neuroscience is the concept of the neuron as a living cell that forms a network, the nervous system. Comparing a single-cell lifeform with a neuron, the former swims in a chemical soup (in the sea, in a drop of water) seeking food and lives to reproduce (by splitting). The neuron is fixed in the brain and has servant cells to feed it (glial cells), and it survives, thrives, or withers and dies according to how useful it is to the brain. A neuron that has few connections and rarely fires action potentials will die. Some neurons, such as pyramidal neurons, act as network hubs and have powerful influence.

The competition between neurons to stay relevant and not whither may also be seen as an act of survival and competition between them (Eagleman, 2021). After all, these are living cells and cells, as Bray relates, have sophisticated survival mechanisms. The chemicals that drive neurons to form synaptic connections are proteins called neurotrophins and neurons that succeed in connecting with neighbors receive these chemicals and survive, those neurons that fail to connect die from lack of neurotrophins. The process of neuron death is either by lack of nutrients (necrosis), or cell suicide (apoptosis). Furthermore, neurons also seek to avoid synaptotoxins that kill their synapses when their activity falls below a threshold (Zoubine et al., 1996).

The human brain has some 80–100 billion neurons, and each neuron in the cerebral cortex has some 10,000 synapses. Neuron synapses fall mainly into two types having either excitatory or inhibitory effects on postsynaptic neurons, depending on the types of neurotransmitters released (Migliore and Shepherd, 2005). Some neuroscientists have elevated this feature into a rule named after Nobel-prize winning scientist Henry Hallett Dale: Dale's principle states that most neurons in the brain are either inhibitory or excitatory but not both. According to Clare Howarth and team (2010), excitatory neurons consume nearly four times as much energy as inhibitory neurons.

The brain is living matter whose neurons form a forest that occupies three dimensions, continually reacting to information signals from the senses (Eagleman, 2020). The human body is mapped to neurons retaining spatial proximity so that a homunculus figure can be drawn over the brain's neurons – the size of the limbs and organs of this homunculus reflect the density of nerve endings. The richer the learning and experiences of a brain the more dendrite and synapse connections are formed.

A baby's brain starts incomplete, some 50% of the neurons are superfluous and die, and there is continual neuronal rewiring as the baby interacts with the world around it. The best analogy of a baby's brain is to think of a sculptor starting with a block of marble and chipping away to create the final figure, similarly the baby's brain starts with a huge mass of neurons that are then whittled into the final shape. The multitude of neurons wire up to the genes' blueprint and while some neurons succeed to make the necessary connections, those that don't die, the redundancy in the initial overabundance of neurons ensures that the necessary connections are created.

The brain is adaptable, where damage occurs other neurons take over: moreover, neurons compete to takeover where other neurons die, for example, due to a limb being lost and its connected neurons have lost the sensory input. Neurons communicate through electrical spikes that travel down their axons (and dendrites see below); these can be triggered by any of the senses, which makes it easy for neurons to take over the activity other neurons, growing new connections, irrespective of which type of sense originated the spike.

Related to this, Eagleman and Vaughn (2021) propose an intriguing theory on the origins of REM sleep, based on research (Merabet et al., 2006) showing neural plasticity operates on a timescale around 90min. Lofti Merabet et al. (2006) and team merely blindfolded their experimental subjects and neurons connected to the visual system were being taken over within 90min! David Eagleman and Don Vaughn therefore point out that a night's sleep can have a deleterious impact on the brain's visual system – unless that system is kept active by dreams that occupy those visual neurons, so spikes travel to the visual cortex to keep it working. To stop the body physically moving to the action in the dream, the brain paralyzes the major muscles groups during dreaming (known as REM atonia).

The brain is continually firing messages even in the absence of intentional sensory input (Greicius, 2003; Kenet et al., 2003; Raichle, 2009), described by Stanislas Dehaene (2014) as the restless brain. In this default brain state, the body is resting with eyes either closed or open but the mind is free thinking or daydreaming; called default mode network and found to largely overlap with the language network (Mineroff et al., 2018; Barrett, 2021). As Raichle points out, debate about intrinsic brain activity has a history in neuroscience and modern functional brain imaging and EEG measurements have provided new evidence for this activity. Kenet and team describe spontaneous brain activity "...dynamically switching cortical states could represent the brain's internal context, and therefore reflect or influence memory, perception and behavior". The key point is that these are not random activities but patterns that resemble responses to visual stimuli (Ringach, 2003).

The restless brain consumes as much as about 95% of the brain's energy (Raichle, 2010), with only about 5% of energy expended on changes of brain state, so there must be a useful purpose for this activity. Dehaene points out that this background activity throughout both hemispheres shows up as a massive, dominating amount of high-frequency electrical waves in EEG measurements, and that neural activation associated with external stimulus is so faint against this background that noise reduction techniques are required to pick out the signals of interest. Recent work compared spontaneous activity in the visual cortex to stimuli evoked patterns and found correlation that supports internal triggered visual thinking or imagination (Kim et al., 2019).

There is a current paradigm that talks of the brain as a prediction machine (e.g., Howhy, 2013; Clark, 2015), and Artur Luczak and team (Luczak et al., 2022) have proposed a neuron learning rule based on localized learning. They suggest that spontaneous activity provides the training data (to use an engineering term) for neurons to build a predictive model. Another important point is that this spontaneous activity continues in our sleep – the brain does not shut down during sleep.

The default neural network is reinforcing patterns that relate to the real world so that our internal thoughts are subjected to a Darwinian selection process (Berkes et al., 2011; Dehaene, 2014). A key aspect to this internal activity is the ability of neurons to fire without external stimulus. For example, a class of hippocampal neurons fires spontaneously without any signaling from senses or other neurons (Kandel, 2006).

Eagleman points out that the brain's adaptability to map sensor information into an internal model of the body, the homunculus, allows it equally to adapt to new limbs and sensors: the brain will extract whatever meaning it can discover from signals it is fed and control in feedback (see Chapter 7 on neurorobotics). From an evolutionary viewpoint, it allows nature to try new body structures, and for scientists today research into brain-machine interfaces is leading to neural control of prosthetics (Lebedev and Nicolelis, 2017; Eden et al., 2022). This adaptability also opens possibilities for new qualia to be sensed by humans, such as vision beyond the visible spectrum, extending into ultraviolet and infra-red for example. Also in animals: a "fish operated vehicle" is reported by Shachar Givon and team (Givon et al., 2022, see also the video referenced, which shows a goldfish driving its tank around a gym court to points where it is fed).

Neurons fire, Eagleman suggests, only when necessary to minimize energy consumption, in the sense that the brain's neural network is configured to predict events around it and a pattern stream that is predictable causes little activity in the brain. In contrast a surprise event causes neuron activity to increase in response to processing new unpredicted information.

Simon Laughlin and Terrence Sejnowski (2003) point out that the brain's communication system follows design rules to optimize resource allocation that are similar to how engineers build electronic communication networks.

The gap junction or electrical synapse is found throughout the nervous system including the brain and is less common than the chemical synapse, which communicates signals via neurotransmitters. Whereas the electrical synapse has direct electrical coupling of potentials between the pre- and postsynaptic points, it also allows exchange of chemicals (Bennett and Zukin, 2004). An important distinction between these types of synapses is that the electrical one can transmit an action potential almost instantaneously, whereas the chemical synapse endures a delay of around in the order of milliseconds (Dong et al., 2018). Gap junctions are also expressed in glial cell types and connect glia and neurons.

Theodore Bullock and team (2005) argue that the neuron doctrine needs expanding to embrace a more complex view of the channels of information processing, from principally neurons to "intercellular communication by gap junctions, slow electrical potentials, action potentials initiated in dendrites, neuro-modulatory effects, extra-synaptic release of neurotransmitters, and information flow between neurons and glia". These additional channels are explored in the following sections.

The neocortex, which largely controls vision, language, and decision-making, comprises about 70%–80% excitatory neurons, the rest being inhibitory interneurons (Tan and Shi, 2013) (interneurons are neurons that sit between sensory and motor neurons).

3.3 ACTION POTENTIAL

The widely understood behavior of the action potential is that it is a voltage change in a neuron that gives rise to an electrical impulse or spike that travels along the axon to its synapses and gives rise to a signal transfer to connected neurons. This statement is unpacked in what follows. Given such an important function, i.e., communication between neurons, there is a surprising lack of detail on this topic outside specialist literature. Research in the last decade has uncovered some fundamental aspects of this phenomena and these findings may simply not have spread to the broader literature.

The action potential fires when a potential threshold is reached, creating a depolarization at the point of initiation, which research consensus points to the axon initial segment (Leterrier, 2018), the initial unmyelinated section of the axon and some 50 μm from the neuron soma, a region dense with ion channels. The threshold voltage level is a common property across neurons, and once triggered has a fixed amplitude so there is no information carried by the size of the spike, however highly stimulated neurons fire with a high frequency of spikes, so the frequency of firing carries information, together with the presence of the spikes (in contrast with no spike).

Campenot (2016) points out that ion channels can be thought of as electric wires across the cell membrane: in the potassium channel, the channel membranes are negatively charged while the potassium ions are positively charged mobile particles. This compares with a metal wire, where the polarities are reversed: the protons form fixed positive charges, and electrons are mobile negative charges carrying current.

To understand action potentials, the first point to note is that the presence of voltage-gated sodium and voltage-gated potassium channels create the ion transfer that gives rise to the spike. The 1963 Nobel prize in Physiology or Medicine was awarded to Sir John Eccles, Alan Hodgkin, and Andrew Huxley for their work in the late 1940s and early 1950s in demonstrating how action potentials arise. The first action potential was measured by Hodgkin and Huxley in 1939. The history of Hodgkin and Huxley's research into the workings of the action potential is detailed by Robert Campenot in his book *Animal Electricity* (2016). Campenot points out that the action potential while typically a spike in the brain lasting about 3 ms can be an extended wave of a third of a meter long when traveling in fast-conducting motor axons that connect the spinal cord to skeletal muscle.

To return to unpacking the first statement of this section, what is less widely known is that the action potential is a wave that will travel in all directions free for it to run, forwards along the axon to its synapses, but also backward toward the dendritic tree – see Figure 3.1. This backpropagation was first discovered by Greg Stuart and Bert Sackmann in 1994 and described in greater detail by Stuart et al. in 1997. This neuron behavior sharing the same name as the training algorithm in artificial neural networks is suggestive and Stuart and Sackmann explore how backpropagation could modulate computational properties of neurons, for example, resulting in the induction of long-term changes in synaptic strength. This role in synaptic plasticity has similarities with algorithmic backpropagation.

During the action potential event, the transfer of sodium and potassium ions is tiny compared to the surrounding sea of ions, but over time, the imbalance could become noticeable, however the sodium-potassium pump, an enzyme, restores equilibrium by transporting ions through the axon membrane. This housekeeping function, as Michael Forrest (2014) describes the accepted view, may need reappraisal as new research on cerebellar

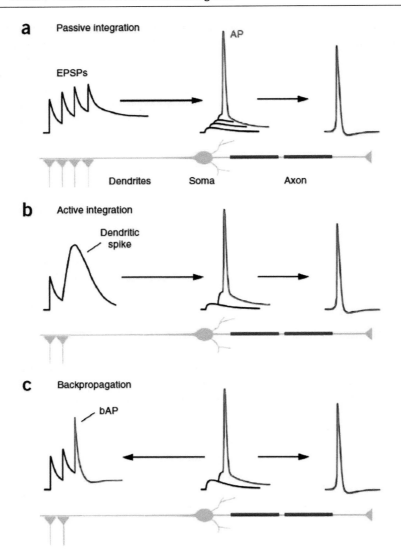

Figure 3.1 Action potential (AP) initiation, dendritic spikes, and backpropagation. (a) Passive integration. Dendritic synaptic input generates fast local excitatory postsynaptic potentials (EPSPs) that are filtered and attenuated as they spread to the soma, where they summate to initiate an action potential (AP, red) in the axon initial segment. This AP then propagates down the axon. (b) Active integration. Dendritic synaptic input initiates a local dendritic spike, which spreads to the soma facilitating AP generation. (c) Backpropagation. In some cells, once initiated, APs (red) actively propagate both down the axon and back into the dendritic tree, where they interact with synaptic input.

Source: Greg J. Stuart and Nelson Spruston: *Dendritic integration: 60 years of progress* (2015). With permission. Note: the original figure text reproduced here contains references not quoted here.

Purkinje neurons by Forrest suggests the sodium-potassium pump may have a computational function. Forrest points out that sodium-potassium pumps account for nearly 70% of energy consumption in the brain, indicating an important role. Purkinje neurons are responsible for motor control. Forrest finds that the pump sets the activity mode of the Purkinje neurons to quiescent or spontaneously firing in continuous slow firing, continuous burst, or various combination patterns of these modes. The pump mediates a form of memory, by changing the axon membrane potential, such as switching off spiking and creating a long period of quiescence (many seconds). Forrest finds the pump can act as a spike integrator, memorizing past firing activity to dictate the timing, amplitude, and duration of quiescent periods. This has an inhibitory effect on downstream neurons.

Huxter et al. (2003) find that the hippocampal pyramidal neuron spikes (in animal brains) have both a rate of firing and a phase (in relation to the hippocampal electroencephalogram (EEG) theta rhythm) in their coding messages, and the rate and phase are independent of each other.

The firing of neurons in the brain in response to sensory inputs is sparse, see Olshausen and Field (2004). Neurons are found to be optimized to produce sparse representations. Sparse responses can, for example in vision, fully represent an image using a small number of active neurons. Sparse coding in the brain has benefits (Olshausen and Field, 2004) as follows:

- It allows for increased storage capacity in associative memories.
- Makes the structure in natural signals explicit.
- Represents complex data in a way that is easier to read out at subsequent levels of processing.
- It saves energy.

Firing action potentials is energetically expensive, and the brain finds ways to reduce neural firing without diminishing performance. This leads to sparse coding strategies and there are several competing theoretical models of sparse coding in the field. Beyeler et al. (2019) propose what they call nonnegative sparse coding (NSC), which represents observed data as a superposition of a set of sparsely activated basis functions and enforcing dimensionality reduction on the inputs. The authors find evidence that a range of neuronal responses are described by NSC.

3.4 DENDRITES

Dendrites are connections that grow in a neuron to receive signals from other neurons, positioned opposite a synapse, across the synaptic cleft. The role of dendrites was originally thought to be receivers of synaptic potential signals and channeling these to the soma and axon. Their role in neural information processing is more diverse: above we saw they also receive backpropagated signals, here we find that they can also initiate a dendric action potential (Figure 3.1b).

Dendrites relay potential levels (also called excitatory postsynaptic potentials or EPSPs) through the soma and to the axon initial segment, where a neuron initiates an action potential. For dendrites distant from the soma, this signal decays over the longer lengths it needs to travel but can be reinforced by contributions from other dendrites over time, especially as EPSPs travel slowly.

The review of dendritic integration of synaptic inputs by Stuart and Spruston (2015) highlights the voltage-dependent channels and electrical properties of dendrites that have axon-like behavior. Electrical dendritic spikes can be generated by many types of neurons; the accumulation of signals traveling to the soma can then influence action potential generation. The duration of dendritic spikes can vary from less than 5 ms to over 10 ms. Dendritic spikes have limited travel distance capabilities and therefore their main influence is on action potential initiation in the axon initial segment.

There is an interaction between backpropagating action potential and dendritic spikes which can lead to initiation of action potential bursting. Backpropagating action potentials into the dendritic tree can also *reduce* subsequent dendritic spike generation. The combination of passive and active dendritic behavior corresponds to computational logic such as involving AND, OR, and AND-NOT logic components.

The architecture of dendrites varies with cell types: the Purkinje cerebellum neuron has a blooming of dendritic branches that resembles a leafless tree in winter, while the bipolar retina cell has an axon that travels away from the neuron body and ends with a smaller dendritic outgrowth. Pyramidal neurons have basal dendrites: dendrites that take synapse signals and send them directly into the soma region surrounding the neuron nucleus, as well as apical dendritic trees (Spruston, 2008).

While some synapses connect directly onto the soma of the neuron body, the synapses connecting through dendrites have signals that travel a further distance. This view of the dendrite is further nuanced by the presence of dendritic spines, membrane protrusions that can alter the electrical environment in the dendrite vicinity (Meriney and Fanselow, 2019).

3.5 GLIAL CELLS

There are as many glial cells in the brain as there are neurons, across many animals including humans the ratio is close to one (Herculano-Houzel, 2016). Suzana Herculano-Houzel goes on to show a high correlation between mass of brain structure and number of glial cells, and notes that

> glial cells must be doing something so sensitive, so important in the brain that the way their numbers are added to build brain tissue has remained very much the same over at least the last 300 million years of evolution.

Woo-Hyun Cho et al. (2016) concur on the importance of glial cells for regulating synaptic transmission based on emerging data.

A type of glial cell called astrocyte has been shown to interact with synapses and influence their plasticity (Pascual et al., 2005; Perea et al., 2009). Pascual and team showed that astrocytes control the level of adenosine, a neurotransmitter, in synapses. The term tripartite synapse was coined to describe the pre-synaptic, post-synaptic, and astrocytic components of a synapse (Araque et al., 1999), which trigger the release of chemical transmitters that regulate synaptic plasticity (Semyanov and Verkhratsky, 2021) – see Figure 3.2. It remains to be understood how ubiquitous the

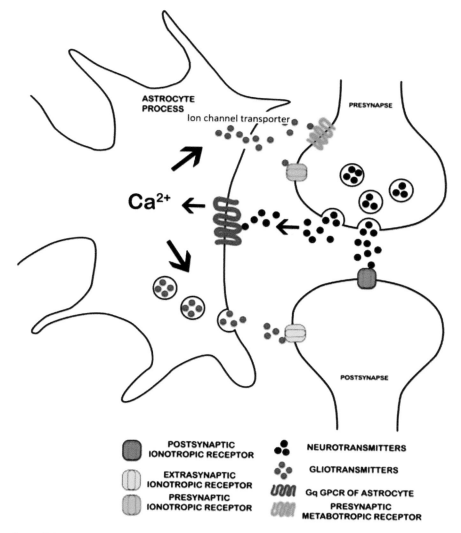

Figure 3.2 Tripartite synapse.

Source: Cho et al. (2016). Ca²⁺ are calcium ions. With Permission.

tripartite synapse is compared with the bipartite synapse. Tetrapartite (four parties) synapse has also been proposed to include the role of the extra-cellular matrix around the synapse (Chelini et al., 2018). Semyanov and Verkhratsky point out that the impact of multi-partite processes at a syn-apse means that signaling between two neurons affects the wider cellular and non-cellular environment.

Perea et al. (2009) make the point that the role of astrocytes is one of an active partner in the integration and processing of synaptic information, controlling its transmission of signals and its plasticity.

Glial cells that myelinate (known as oligodendrocytes cells) can detect action potentials flowing in axons through interacting with binding mol-ecules at axon membrane receptors, thereby creating communicating chan-nels other than through the synapses, propagating communication through the glial network, going beyond the neural network (Bullock et al., 2005). Myelin sheaths grow to full maturity by the time humans become adult, which explains why children think more slowly than adults (axon insula-tion increases spike velocity). Note that the brain tissue in the forebrain is known as "white matter" because it comprises neurons whose axons are sheathed in white myelin. However, in the cerebral cortex, which exists only in mammals and is the outer layer of the forebrain, the neuron axons are mostly unmyelinated (Bermudez, 2023), giving rise to the name "grey matter".

Glial cells communicate by molecular signaling. There are also chemi-cal synapses between neurons and glia known as oligodendrocyte progen-itor cells (OPCs, cells that create oligodendrocytes). Bergles et al. (2000) have shown that there exists "a rapid signaling pathway from pyramidal neurons to OPCs in the mammalian hippocampus that is mediated by excitatory, glutamatergic synapses". Furthermore, glia actively sculp neu-ral circuits during brain development and in the adult brain (Buchanan et al., 2023). Bullock et al. cite evidence for astrocytes forming a com-munication network that runs in parallel with the brain's neural network but at a slower time scale.

3.6 GRID NEURONS

Grid cells are neurons located in the dorsocaudal medial entorhinal cortex (dMEC) region of the brain. This region maps an animal's physical location in the external world to a model of that spatial environment in a topographically organized neural network. These grid neurons activate when the animal moves to the corresponding physical location. According to Hafting et al. (2005), this map is anchored to external landmarks and persists when the landmarks are not visible. The neural network connections have a honeycomb, hexago-nal shape (Buzsáki, 2019). The position of one object to another in the map is allocentric or object-to-object, in contrast with egocentric (object to self).

Grid cells are the most abundant in dMEC, but the region contains other environment-reflecting cells (Rowland et al., 2016):

- *Grid*: location in the environment.
- *Border*: proximity to geometric borders.
- *Speed*: running speed.
- *Head direction*: indicates the orientation of the animal relative to environment landmarks.

These cell classes are clearly distinguishable from one another, and cells never switch from one class to another. Killian et al. (2012) cite evidence from experiments on monkeys that, as a monkey's gaze moves around the environment, grid cells fire up as if the animal was moving through the locations being viewed. The firing of location-related grid cells through thinking of movement is a clue to visual thinking relying on grid cells. Finally, 2D spatial representation is a fundamental property of the brain given that the eyes provide only 2D data. However, we live in a 3D world and Ginosar et al. (2021) have identified dMEC cells that relate to 3D space.

3.7 MIRROR NEURONS

Mirror neurons are a class of neurons that fire when an individual executes a motor action (e.g., picking up an object) and the interest is that they also fire when the individual observes the same or similar action carried out by another individual. These neurons were first reported in macaque monkeys in 1992 and subsequently in the human brain (Kilner and Lemon, 2013). The concept attracted notoriety, culminating in the book by Gregory Hickok, *The myth of mirror neurons* (2014), which does not dispute their existence but rather casts doubt on dubious theories (the myths of the title) that flourished on the back of the discovery. David Kemmerer (2015), in his paper on the topic, the sub-title of which states it to be reflections on Hickok's book, praises Hickok for calling out on the hype, agreeing that "mirror neurons are not the fundamental 'basis' of action understanding", nevertheless, stresses that motor system do play a significant role in the perception and interpretation of actions. Mirror neurons have been tied up with theories of language evolution in the human brain, e.g., see Michael Arbib's (2012) book *How the Brain Got Language: The Mirror System Hypothesis*.

3.8 ELECTRICAL COMMUNICATION IN THE BRAIN: AXONS, DENDRITES, NUCLEUS, AND SYNAPSES

The brain is an electric machine as well as a chemical machine. We explore some aspects of this electric nature beyond the action potential discussed above.

Donald Hebb (1949) suggested that if a cell fires and causes a connected cell to fire as well on a consistent basis then the connection between them increases, and there is a metabolic or process growth at their synapse/dendrite junction. A neuron firing will spread its spike energy to all its synapses, so many connected neurons are being influenced in parallel: according to how the connected neighboring neuron is influenced (i.e., whether they fire or not), their junction strength increases or decreases, but the timing is crucial, connections are strengthened when a spike triggers an immediate spike in a connected neuron. This led to the Hebb rule: neurons that fire together, wire together. There are chemicals that detect coincidence of this firing influence at the synapse junction supporting Hebb's idea (Eagleman, 2020).

Most non-neuroscientist readers I expect are familiar with the synapse as a one-way channel junction, a rectifier that receives a spike signal, converts it into a chemical transmission process in the junction which then activates a potential rise in the dendrite of the post-synaptic or receiving neuron. However, there are also present purely electrical synapses in human and other animal brains. Moreover, these junctions are typically two ways, with strength variations depending on traffic direction (Pereda, 2014; Faber and Pereda, 2018). As Barry Connors and Michael Long (2004) say "Electrical synapses are a ubiquitous yet underappreciated feature of neural circuits in the mammalian brain". Electrical communication came after chemical communication in the evolution of life and exists in two forms: electrical synapses and inter-neuronal field effects (Weiss and Faber, 2010), also known as ephaptic coupling – see below. An electrical synapse has a gap junction comprising clusters of channels that allow transfer of charged ions (Bennett and Zukin, 2004). Figure 3.3 illustrates the three typers of synapse junctions.

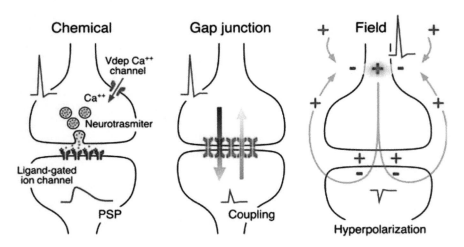

Figure 3.3 Chemical and electrical mediated mechanisms of synapse communication.

Source: Faber and Pereda (2018). Ca⁺⁺ is calcium ion. With permission.

Each of these synapse junction types has its different advantages. A chemical synapse can amplify a presynaptic signal, and as mentioned acts in one direction. An electrical synapse transmits faster than the chemical junction, is bi-directional and can synchronize activity in a neuron cluster. An electrical synapse is also more reliable with less "moving parts" than a probabilistically driven chemical junction (Pereda, 2014). Depending on the polarity of the ions crossing the gap, electrical signaling can act to excite or inhibit adjacent neuron activity. Both types of synapse, chemical and electrical, can be modified by neuromodulators, and have plastic responses to activity. Pereda (2014) argues for the combined processes of chemical and electrical communication in the typical transmission of signal in a synapse.

A spike signal is generated as the result of ions flowing in channels in the neuron axon membrane. The spike signal, caused by ions creating an action potential (see section above), a sudden peak in voltage change in the axon membrane, is an all-or-nothing event: once the threshold for generating the spike is reached, the signal itself is identical in shape and amplitude in all instances. This begs two questions: in conveying information downstream, how does a neuron differentiate different stimulus intensity and different types of stimuli? For answering the former, Edgar Adrian (Nobel prize lecture, 1932; Kandel, 2006) found that the *frequency* of action potential emissions corresponds to the intensity of the stimulus. Furthermore, the duration of action potential generation corresponds to the sensation duration. For the latter question, neurons treat all sensory information identically, information on what type of sense is involved is in the context of the neuron's topological location.

Synapses are dynamic, their connection strengths vary on a time scale from milliseconds to days, weeks or longer (Zador and Dobrunz, 1997) and with large-scale changes to those strengths. This dynamism occurs during "normal" operations in contrast with artificial neural networks where connection weights are altered only during training and remain static during inferencing (Maass and Zador, 1999; Natschläger et al., 2001).

The third neuron-to-neuron coupling is due to their electric fields and is called ephaptic interactions; they can occur through the electric potential fields emanating from neurons affecting neighbors in tightly packed axon geometries. Sheheitli and Jirsa (2020) simulated the effects of coupling in an axon bundle and found that ephaptic interactions can lead to phase locking between adjacent spikes. As the coupling is increased, spikes trigger new spikes along adjacent axons. Further increase in coupling leads to spikes traveling laterally and backward, creating complex spatiotemporal patterns. They conclude that rather than modeling axons as mere relays of signals, the axons as active media should be considered.

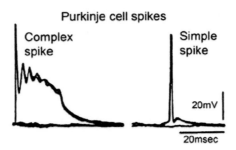

Figure 3.4 Contrast between complex and simple spikes.
Source: Llinas (2013). With permission.

Anastassiou and Koch (2015) note that ephaptic coupling effectively injects intracellular currents, altering membrane potential. Also, changes in the extracellular potential caused by extracellular fields influence synaptic currents. And there are also effects in the synapse cleft caused by intrasynaptic electric fields.

"Climbing fiber" Purkinje cells have some unusual properties (Han et al., 2020): they lie in a single layer, and their dendrites are parallel to one another, like sea fans in a coral reef. Being densely packed, their somas are spaced only a few microns apart. They are powerful neurons in the central nervous system, firing complex spikes and simple spikes (Llinas, 2013; Streng et al., 2018). Llinas (2013) addresses spontaneous neural activity in the inferior olive and cerebellum, including intrinsic brain oscillations (see Section 3.16). Inferior olive neurons reach into the cerebellum with excitatory synaptic contacts and connect with branches of "climbing fiber" Purkinje neurons. Purkinje cells when activated fire both low-frequency complex spikes and high-frequency simple spikes, see Figure 3.4: these cells are the only output of the cerebellar cortex and form inhibitory synapses. Streng et al. (2018) believe complex spikes provide predictive signals.

Han et al. (2020) found that a spike from the Purkinje cell can suppress neighboring spikes for several milliseconds through ephaptic coupling. Ephaptic coupling is a third way by which neurons communicate with each other, after neurotransmitter crossing a synapse and gap junctions where direct electric potentials cross, ephaptic coupling arises from the electric fields produced by neurons, and if strong enough they can instantly excite neighboring neurons. The authors found that a single inferior olive climbing fiber input could through ephaptic coupling impact 18 neighboring Purkinje cells, pausing their firing for several milliseconds. They deduce that a single inferior olive spike could affect 100 Purkinje cells, affecting activity in the forebrain.

3.9 MOLECULAR COMMUNICATION IN THE BRAIN: NEUROMODULATORS AND NEUROTRANSMITTERS

Neuroscientists are exploring the role of neurotransmitters acting in volume as a component of information transfer beyond synaptic transmission, working in a different way from electrical transmission. This chemical transmission is slower acting with a wide reach through the brain (Taber and Hurley, 2014).

Information is transmitted in the brain by electric signals and by molecular (or chemical) interactions, the former being wired via myelin insulated axons, the latter, sometimes fancifully referred to as wireless communication, are transmitted by diffusion through concentration gradient and through thermal agitation, also called Brownian motion. Cells are complex molecular factories, and two molecules that need to meet to complete a process step will do so within a short time span: a small molecule such as a sugar can cross a human cell from side to side in one tenth of a second, a larger molecule such as a protein would take a second, inside the soup of molecules in the cell meeting other molecules along the way with a high probability of meeting the right counterparty necessary to complete a process.

Within a neuron, molecular transmission can be aided, for example, in dendrites and axons, by microtubules (Baas et al., 2016). Microtubules provide structural support inside a neuron, but they also act as molecular railways – little "hands" on the microtubule pass along a protein or organelle to the far end of an axon or dendrite.

Characteristics of neural communication are constrained by the energy expended to create electrical spikes (consuming a large part of the available energy) and the need to maintain signal over noise (Sterling and Laughlin, 2015). This leads to the use of concentrated spike bursts, whereas over short distances of nanometers to micrometers, when diffusion acts effectively, and time frames from 100 μs to s, Sterling and Laughlin argue chemical processing is orders of magnitude less energy consumptive. Computing with proteins in the mammalian brain offers as many as 50,000–80,000 different protein options allowing rich behavior. However, to convey a chemical signal over longer distances, such as to reach an axon end a meter away in a limb and reach it at the same time scale as the chemical reaction requires switching to electrical transmission. A spike can travel 50 times faster in a dendrite than chemical diffusion and 1,000 faster in an axon than chemical diffusion (Sterling and Laughlin, 2015). The neuron therefore serves to convert signaling at the protein level to electrical signals that can transmit signals to other neurons over larger distances.

Edelman and Tononi (2000), in their overview of the brain's neuroanatomy, describe a topological arrangement of a diffuse set of connections that fan out across the brain and originate from a small set of neurons concentrated in the brainstem and hypothalamus that release neuromodulator chemicals, of which there are seven key ones:

- *Acetylcholine*, has an excitatory effect.
- *Adrenaline*, also known as *epinephrine*. *Noradrenaline* is continuously released at low levels, adrenaline is released during stress, triggers flight or fight response.
- *Dopamine*, affects feelings, motivation, reward.
- *GABA*, has an inhibitory effect.
- *Glutamate*, has an excitatory effect.
- *Histamine*, regulates sleep and cognitive functions.
- *Serotonin*, carries messages between cells.

The term neuromodulation describes the process of a neuron using possibly any number of neuromodulators, to affect other neurons, and neuromodulators can be neurotransmitters, neuropeptides, and hormones that over an area and a period impact the synapses and neural circuitry. Neuromodulators force neurons and their synapses to reconfigure according to contextual demand (Mei et al., 2022). The brain can fire any number of neuromodulators to force a change across neurons. They can impact synaptic plasticity, theta oscillations, and neurogenesis. Neuromodulators are mostly emitted from sub-cortical regions of the brain.

For example (Edelman and Tononi, 2000), the noradrenergic locus coeruleus consists of a few thousand brainstem neurons that send out a diffuse "hairnet" of fibers across the brain influencing billions of neurons. These neurons fire whenever important events occur such as loud noises, flashes of light, and sudden pain. When these neurons fire they release by diffusion neuromodulators that influence neural activity. While there is no consensus as to their function (Edelman and Tononi designate them as value systems), their malfunction, even to a small degree, is associated with severe mental illness. Under normal functioning, neuromodulators influence changes in neurons such as plasticity. Synapses can be in a state of reception so that when a neuromodulator is broadcast those synapses can make use of the chemical.

One of the challenges in cognitive learning is how to reward right action in a long sequence of actions, known as the credit assignment problem. Dopamine is the brain's reward neurotransmitter; its release creates a feeling of pleasure and reinforces the brain to repeat action that leads to its release. The question is then how does the brain release dopamine and solve the credit assignment problem. A recent study by Jonathan Tang et al. (2023), a team led by Rui Costa, CEO of the Allen Institute, addressed this question. In collaboration with other research labs, the authors created a closed loop system in mice to connect their specific actions to dopamine release. The mice were implanted with wireless sensors to monitor their movement and used optogenetics (controlling neurons with light) to stimulate dopamine releasing neurons whenever the mice executed the target actions (arbitrarily selected by the researchers). Very quickly the mice reformed their behavior to receive dopamine, focusing on the right action at the time of the hit as well as reinforcing the actions just prior to the hit. Thus, the way the brain uses dopamine is not just a simple reward but as a way of actively shaping and fine-tuning behavior through experience, and the experimenters report that the mice performed this learning process rapidly.

The experimenters then changed the target actions so that there was a longer time interval between the rewards. This resulted in the mice's learning process taking longer to form correct behavior, with actions near to the hit receiving immediate reinforcements, while more distal behavior took longer to establish the correct pattern of action, requiring exploratory, trial-and-error behavior to discover the optimal behavior. The key takeaway is that the mouse uses exploratory behavior to hone on (and learn) the optimal behavior that maximizes dopamine hits.

The role of acetylcholine and noradrenaline has been clarified by Yu and Dayan (2005), who propose that acetylcholine is released when there is expected uncertainty (there is a known unknown) while noradrenaline is released when there is unexpected uncertainty (unknown unknown). Based on existing data in the field, the authors believe these uncertainty signals work in combination to perform "optimal inference and learning in noisy and changeable environments".

Research on how neuromodulators affect spike-timing dependent plasticity (STDP) is reviewed by Brzosko et al. (2019), where the authors conjecture that neuromodulators bridge the time gap between millisecond STDP processes and the slower forming behavioral learning.

3.10 HUMAN MEMORY

Human memory is unlike computer memory. Computer memory is localized in specific storage units such as registers, caches, solid state drives, and rotating hard drives, some volatile and some non-volatile (retain memory when power is switched off), and provides exact recordings of the original information. In artificial neural networks memory is distributed in the weights assigned to connections between network nodes, but these memories are quite brittle and apt to be overwritten and "forgotten" if a new and different application is used to re-train the network. In contrast, human memory has a great capacity to continually learn and memorize new information throughout the lifetime of an individual, building on top of what already exists.

In what follows, some elements of memory research are related. There is still work to be done to fully explain how human memory works but much of the basic machinery has been elucidated, the gaps are mainly at the higher levels of memory, how local memory trails (neurons hold memory) and their memory fragments (or contributions) relate to the processing centers in other regions of the brain and integration of relevant memory fragments into what we experience as a memory, a full "cinematic" internal view. This final step involves recreation rather than retrieval (the fundamental difference between our wetware memory and software memory).

There are three high-level categories of human memories: sensory memory, short-term memory, and long-term memory, see the reviews of the scientific literature by Squire (2004), Kandel (2006), Squire and Kandel (2009), Hasselmo (2013), Kandel et al. (2014), Camina and Guell (2017) – in Camina and Guell there is a useful diagram of the types of human memory, see Figure 3.5. Eric Kandel's scientific autobiography (2006) provides insightful background into the molecular processes underlying memory; for

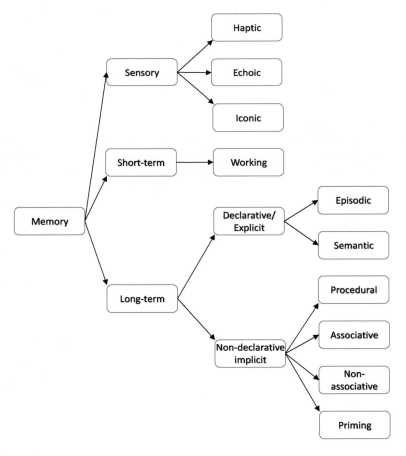

Figure 3.5 Classification of the various types of memory in the human brain.

Source: Based on Camina and Guell (2017). With permission.

his work on the sea slug Aplysia he won a Nobel prize in 2000 (the book has a nice photo of the medal hung around a sea slug's neck).

Short-term memory is held by synapses but as explained in Kandel et al. (2014) so is long-term memory. Kandel and his team found that long-term memory, in Aplysia but also memory storage in all animals, follows three principles (quotes from Kandel, 2006):

- "Activating long-term memory requires the switching on of genes". The neuron nucleus holds genes which "express" proteins essential for the memory molecular processes. However, note that the latest research (Hafner et al., 2019) shows that synapses can also produce proteins locally. A single neuron can contain about 50 billion proteins which regulate synaptic plasticity.
- "There is a biological constraint on what experiences get stored in memory". Genes that switch on for long-term memory are accompanied by genes switching on memory suppression, ensuring a high threshold for converting short-term to long-term memory – most experiences are forgotten. The removal of the suppressant allows long-term memory.
- "The growth and maintenance of new synaptic terminals makes memory persist". Laying long-term memory grows new synapses.

The question of how long-term memory is maintained over time against protein degradation (proteins have a limited lifetime of about a week) was answered by Kandel and his co-worker Kausik Si (Si et al., 2010) who found a role for prions, which hitherto were associated with neurodegenerative diseases such as mad cow disease. The prion's role is to perpetuate synaptic memory storage and to do so selectively at a single synapse out of the possibly thousands of a neuron's synapses. Long-term memory is synapse-specific, and signals are sent from a specific synapse back to the nucleus and from the nucleus to specific synapses – there is a two-way chemical communication taking place. The nucleus broadcasts the products of its gene expression to all its synapses, but only the target synapses receive it by being "tagged" by previous activity. The tag is CPEB, an active and non-pathogenic prion. Long-term memory is therefore dynamic and must be continually refreshed by the brain.

Sensory memory is linked to the point where sensory information enters the body, and the three main senses each have an associated memory: iconic for vision, echoic for sound, and haptic memory for touch. There is a lot of research on iconic memory, it is found to have a short span of about one second, forming a reservoir that feeds into short-term vision memory. Long-term memory stores information for long periods of time (certain such memories will last a lifetime) and retrieval of this memory can be performed consciously (explicit memory) or unconsciously (implicit memory). Explicit memory, also called declarative memory, stores facts, events, people, places, and objects.

Working memory (Baddeley, 2010) provides temporary storage for information manipulation, requiring complex processes such as understanding language, learning, and reasoning. There is a phonological buffer (where new words are temporarily stored and linked to articulation rehearsal) found in the left brain.

Explicit memory calls on activity in the hippocampus and adjacent cortex and involves conscious awareness; in contrast, implicit memory is an unconscious process and calls on activity in the cerebellum, striatum, and amygdala. In invertebrate animals, implicit memory is located in the reflex pathways.

Sensory neuron to motor neuron synapses that mediate reflex actions also underpin learning and memory (Kandel et al., 2014). Storage of implicit memory is built into the brain's neural networks and is facilitated by synaptic plasticity. The encoding and storage of short-term memory takes place between the sensory and motor neurons controlling muscle movement. Kandel notes that "the cellular mechanisms of learning and memory reside not in the neuron but in the connections it receives and makes with other cells in the neural circuit to which it belongs". And further that

> even though a neuron may make one thousand or more synaptic connections with different target cells, the individual synapses can be modified independently in long-term as well as short term memory. The synapses' independence of long-term action gives the neuron extraordinary computational flexibility.

Stimulation of a sense releases the neurotransmitter serotonin onto the sensory neurons, and the synaptic connection between the sense and motor neurons is strengthened. The effect of serotonin is to increase the concentration of cyclic adenosine monophosphate (cAMP) in the sensory cell, and cAMP molecules signal the sensory neuron to release the neurotransmitter glutamate into the synaptic cleft, strengthening the connection between the sensory neuron and motor neuron. The sequence of events starting with serotonin results in a catalytic molecule moving into the neuron nucleus.

An important function in the brain is the transfer of short-term memory to long-term memory locations in the brain (Pasupathy and Miller, 2005), where skills learned by repeated actions become habituated. Much about the workings of memory were learned by studying a patient known as HM (Squire, 2009) who had server lesions in his brain.

Eagleman (2020) argues that synaptic plasticity while necessary is unlikely to be sufficient to encode memory. For example, the cortical changes found in London taxi drivers who navigate the streets by memory resulted in more than synaptic modifications, brain images showed cortical structure changes (Maguire et al., 2000: "The posterior hippocampi of taxi drivers were significantly larger relative to those of control subjects"). The growth of neurons (neurogenesis) explains such changes (Boldrini et al., 2018).

There is also the impact of epigenetics (see Section 8.7.3), where the experiences a person encounters affect the DNA in the neuron nucleus by magnifying some genes and shrinking others (Levenson and Sweat, 2005; Lisman et al., 2018). David Eagleman argues that multiple processes and parameters beyond spiking are at play, operating at differing time scales, to drive synaptic plasticity, such as neuron membrane ion channel types and distributions, phosphorylation states, growth of neurites (which may form new dendrites), rate of ion transportation, and gene expression (through epigenetics). Learning and memory therefore involve multiple strands of brain plasticity (Eagleman, 2020).

3.11 BRAIN LATERALIZATION

The left and right hemispherical split in the brain, also known as brain lateralization or brain asymmetry, is now understood to occur in most if not all vertebrates (Rogers et al., 2013b) and in some invertebrates as well (Frasnelli, 2013). The question for HLAI researchers is whether lateralization is a requirement of intelligence, or an evolutionary feature retained for other beneficial reasons. For example, some marine mammals and birds sleep with one hemisphere asleep while the other is alert (Mascetti, 2016). Given the near-universal ubiquity of lateralization in the animal kingdom, there may be a link with intelligence – this is discussed further in Part 3, but here we focus on what type of asymmetries exist in the brain.

The left and right hemispheres are connected by a dense bundle of connecting axons, some 200 million, called the corpus callosum, which transfers information between the hemispheres. To put in context, the information carried by this axon bundle is a very small fraction of the information processed in the vast network within each hemisphere, which implies that much work performed in each hemisphere is independent of the other side but there is information sharing across the sides.

Following the most recent research in brain lateralization (Corballis, 2003, 2014; Rogers et al., 2004; Rogers et al., 2013a; Ocklenburg and Gunturkun, 2017; Rogers and Vallortigara, 2017), the following is known about human brain lateralization (and these findings pertain to the majority of people, there will be exceptions, and some characteristics are human age related):

- The brain hemispheres divide workloads, there is no case where only one hemisphere processes some workload and nothing is performed in the other half – the two sides are in continual communication; however, each side processes different aspects of the workload. In language comprehension, Kara Federmeier's research indicates the left side performs top-down brain predictive processing while the right side performs bottom-up feedforward processing (Federmeier and Benjamin, 2005; Federmeier, 2007; Federmeier et al, 2008).

- The left hemisphere is slightly larger than the right in most people.
- The "seat of language" is principally in the left hemisphere, but not wholly, both sides are necessary for language processing. While the left side governs speech the right side does comprehend language while not involved in the verbalization process. The left side is also dominant for sign language (Pettito et al., 2000).
- "Handedness" (or equivalent "footedness" in animals such as parrots), is largely governed by the left hemisphere (so you are likely to be right-handed because of the cross-wiring of the brain).
- In performing arithmetical tasks, both sides of the brain are involved but verbal representations and phonological output is governed by the left side (Dehaene et al., 2004).
- The right hemisphere receives sensory inputs first, before relaying to the left side. Each side governs movement on the corresponding opposite side (because of cross-wiring).
- The right hemisphere is associated with emotions.
- The sense of self-awareness and recognizing that other people have a similar mental life (known as theory of mind) is associated with the right hemisphere.
- Processing spatial relations is associated with the right hemisphere.

There is another intriguing aspect to brain lateralization. Which has to do with the cross-wiring that features in all animal brains from humans to nematode worms: body sensors on the left side of the body connect to the right side of the brain and right-side body sensors connect to the left-brain hemisphere. Research by Troy Shinbrot and Wise Young (2008) puts forward the idea that the brain cross wires for optimal topological mapping of 2D sensory input onto the brain's 3D structure of the world. Decussation is what neuroanatomists refer to the crossing over of axons between motor and body sensors and the hemispheres, common across brains in the animal kingdom. Understanding the reasons and benefits of decussation led Shinbrot and Young to their research and the mathematical hypothesis behind the mind's 3D rendering of 2D sensory data.

The needs of mapping the external world in the brain (Land, 2014) may have led to an initial left-right split of the brain hemispheres. Note that the way the eyes connect to the hemispheres is not a simple crosswire: the right half of each retina is connected to the right hemisphere and the left halves connect to the left hemisphere (Hoffman, 1998).

While the left-right brain split is a physical attribute of the brain, there is also debate in the neuroscience field into what is called a two-stream model between dorsal (top) and ventral (bottom) processing of vision and more (Goodale and Milner, 1992), that is a processing path split and not a physical feature of the brain. An assessment of the theory and its evidence has been given by McIntosh and Schenk (2009).

Finally, the question of what happens if the corpus callosum is missing in the brain, the case of Kim Peek is of interest. Peek was an "idiot savant" and the model for the fictional Rain Man story who was missing the anterior commissure and the hippocampal commissure, both regions providing connections between the hemispheres, as well as missing the corpus callosum, the main connecting network between the hemispheres (Brogaard, 2012). Peek was able to read both pages of a book simultaneously, one eye on each page and had phenomenal recall of information gleaned from reading. In normal brains information from the left eye is channeled to the right hemisphere and is then directed to the left hemisphere, the main center for language, via the corpus callosum. Without this transfer process, Peek developed language abilities in the right hemisphere which gave him great powers of information absorption and speedy processing.

While Peek's case is of a natural brain impediment, there is a surgical operation severing the corpus callosum performed to reduce seizures in epileptics. The resulting split-brain personality leads to conflicting actions as each hemisphere tries to control the body. Peek had a low IQ due to the lack of certain thinking capabilities associated with the hemispheres communicating. This is a clue to the link of lateralization to intelligence.

3.12 BRAIN FOLDS AND NEOCORTEX COLUMNAR STRUCTURE

The human brain structure is highly folded in concertina manner (gyrencephalic, see Garcia et al., 2018). Jeff Hawkins (2021) describes the neocortex (which occupies over 70% of the human brain) when stretched out as being "large napkin sized", around $1 \text{ m} \times 1 \text{ m} \times 2.5 \text{ mm}$ with a regular structure, although the uniformity across species is a simplification. A highly cited paper by Rockel et al. (1980) first proposed uniformity across species, who stated that in different animals the absolute number of brain cells through the thickness of the cortex was constant. Uniformity within the brain and across species is disputed by Guy Elston (2003), who finds that "pyramidal cells in the prefrontal cortex have, on average, up to 23 times more dendritic spines than those in the primary visual area". Non-uniformity is supported by Herculano-Houzel et al. (2008) and Rakic (2008), while Carlo and Stevens (2013) verify the uniformity proposal.

Starting with Mountcastle in 1957 (see Mountcastle, 1998) the neocortex is understood to have cortical columns containing repeating groups of neurons across six layers. Rakic points out that other of Rockel's conclusions are not in dispute, that the neocortex has a basic columnar structure and through evolution it spreads in surface area rather than in layer thickness.

The advantage of folding, or cortical gyrification, is to create greater surface area in a fixed volume. The key consequence of this is that it allows a greater number of neurons to populate the volume of brain and explains the evolutionary benefit of gyrification. While neurons under bump surfaces (the gyrus) are brought slightly closer to each other, gyrification has little change in axon distance traveled between widely spaced connected neurons (axons do not tunnel through surfaces to bypass the neocortex troughs, the sulcus).

From an evolutionary perspective, Striedter and Nothcutt (2020) examined gyrification in six major mammalian lineages. In all cases, the smaller animal members of each lineage had a smooth neocortex compared with larger-sized members. For example, in the Laurasiatheria group the horseshoe bat has smooth neocortex compared with the gyrencephalic dolphin, and in the Euarchontoglires group the house mouse is smooth versus the gyrification of the chimpanzee. Gyrification therefore correlates* with greater animal intelligence. This has no implications for HLAI where software and hardware are not constrained in the way animal brains are restricted by space and resources, for example, software models can work in a multi-dimensional space. (*Note: as always, correlation does not imply causation.)

A model to explain how gyrification occurs and leads to the brain's shape and structure, its morphogenesis, has been given by David Van Essen, called tension-based morphogenesis (Van Essen, 2020).

3.13 EARLY BRAIN DEVELOPMENT

Human brain development follows a mixture of nature and nurture, i.e., genetic and environmental influences: genetic through blueprints embedded in the DNA and environmental through plasticity of neurons and other "pieces" in the brain "game board" activated by experiences encountered in life. Genes and experiences are also intertwined through epigenetics. Genes express proteins which are akin to encapsulated algorithms. The benefit of being born with a natural blueprint gives the newborn immediate survival traits and behaviors, which can be further developed by brain growth adapting to experiences encountered. The epigenetic process allows experience at a molecular level to affect how the genes express proteins so the in-built behaviors can be modified to affect the animal's actions. This is a dimension of freedom beyond the initial newborn's balance between nature and nurture and all these dimensions may be useful in HLAI.

The human brain development starts in the embryo and continues to grow and adapt into adult life. In the embryonic stage, neurons are produced by progenitor cells, a type of stem cell, in a neural tube, described by Bryan Kolb and Robbin Gibb (2011) as the brain's nursery. The neural tube grows into what becomes the brain including the ventricular zone and the spinal cord. Progenitor cells are among the many transient cells eliminated by the end of early brain development. Most neurogenesis is completed by five months from conception, except in the hippocampus which is notable for neuron production throughout life: Spalding et al. (2013) have shown that some 700 neurons per day are added in each half of the hippocampus (split between the left and right sides of the brain) in adult humans, with a small decline in old age.

In the fetus, some 250,000 neurons are formed per minute, according to Kolb and Gibb. Neurons migrate by moving along radial glial cells that span out from the ventricular and sub-ventricular zones to the cerebral cortex. The plan of migration follows an algorithmic-like genetic blueprint, the precise nature of which is unknown (Koulakov et al., 2022), and the migration process is completed before birth. Once in position the neurons begin to form neural networks (Innocenti and Price, 2005; Stiles and Jernigan, 2010), each neuron growing an axon and an arbor of dendrites. The tips of axons and dendrites have growth cones (so named by Santiago Ramón y Cajal) and as the axon/dendrite elongates the cones sample the environment for guidance molecules which direct the axon/dendrite to its target. Guidance molecules perform attractive or repulsive effects to guide the growth of the axon/dendrite (Stiles and Jernigan). Connections are formed abundantly. In the immediate postnatal period, the connectivity between neurons exceeds that of an adult, and as the brain matures a pruning process takes place, dissolving axons, synapses, and terminating neurons. According to Kolb and Fantie (2008), synapse loss runs to 100,000 per second in adolescence. New neuron growth is then a response to environmental experiences.

After the massive reduction in axons the process of myelination begins, which is the process of electrically insulating axons. The lack of myelin in early development of the brain results in sluggish response times to sensory information and internal thought processes. Myelin, a fatty substance that grows around the axons, through acting as an insulator increases the speed of electrical propagation, reducing signal dissipation and fidelity.

Innocenti and Price (2005) discern between macroscopic exuberance of (what eventually become transient) connections between different brain regions and microscopic exuberance where (again eventually transient) connections occur between neurons in a localized area. Many of these connections do not exist in adults and are removed during brain development. In thalamocortical development, the transient projections are thought to help guide thalamus axons to their targets. Macroscopic exuberance leads to establishment of main axons while microscopic exuberance leads to establishment of axonal branches and synapses near targets – peak synapse formation occurs between one and two years (Kolb and Gibb) and early synapse formation is not generated in response to experience. Furthermore, Innocenti and Price cite evidence that axons compete against each other in early development to survive and establish themselves. Neurons possess a trait to survive by proving themselves useful – they rely on neurotrophic factors released by their target neurons at synapses. Neurons that lack sufficient neurotrophic factors trigger certain genes that instruct those neurons to self-destruct. Synapses are also pruned from active neurons. These are synapses that do not form active connections, and the lack of activity is believed to kill the synapse, the driver being environmental experience. Kolb and Gibb postulate that the uncertainty in the process of neurons reaching their targets and making appropriate connections is the reason for the overproduction of neurons. Neurons at the migration stage have the potential to form a variety of cell types, upon reaching their targets neurons differentiate into certain types through influences in the immediate local environment, such as nearby genes.

The sub-ventricular remains an important zone throughout the life of the individual. It is lined with stem cells that can be activated to produce neurons or glial progenitor cells. Not well understood, Kilb and Gill cite work that suggests these stem cells become active by factors including experience, drugs, hormones, and injury.

Dendrites start growing once neurons have reached their destinations and continue to develop well after birth. The production of dendrites follows an independent and slow process growing from neurons at the pace of micrometers per day, whereas axons grow at a pace of 1 mm per day (Kolb and Gibb, 2011), allowing axons to reach their target neurons before dendrites make connections.

Dendrites grow as individual processes from the neuron body, and then develop into tree-like structures with multiple branches. Dendrites grow spines (described by Kolb and Fantie as like thorns on a rose stem) that form synapses connecting with axons. The spines can form rapidly in response to experiences (in minutes to hours), and this plasticity is a feature of the brain throughout the individual's life (Kolb and Gibb, 2011). Dendrites modify their structure rapidly in response to experience and can create or destroy synapses accordingly, possibly in minutes (Kolb et al., 2017).

Plasticity in the brain in response to experiences tends to be localized in certain spots and not widespread. The plastic changes are also time-dependent, and once the external influences cease, they may revert to a previous state (Kolb and Gibb, 2011). Glial cells (which do not transmit electrical signals) start to proliferate after neurons are produced and continue to be created throughout the life of the individual (Kolb and Fantie, 2008) – they play an important role in chemical transmission for signaling and nutrition.

Finally, the question of the neuron connection anatomy has been investigated and the connection pattern has been likened to a small world network. Small world network connectivity lies mid-way between a regular connectivity pattern (completely predictable) and a random connection pattern. The degree of small world network connectivity occupies a spectrum between the two extremes. A small world network is defined as a network where the typical distance (measured as hops across nodes) between two randomly chosen nodes is proportional to the logarithm of the number of nodes in the network. The term "small world" comes into use because two random strangers are connected by a small number of hops across acquaintances (Watts and Strogatz, 1998).

Henry Markram and his team at the Blue Brain Project, Ecole Polytechnique Fédérale de Lausanne (Perin et al., 2011) investigated synaptic connectivity in groups of pyramidal neurons in the neocortex. They found synaptic weights closely followed the number of connections in a group of neurons, saturating after 20% of possible connections were formed between neurons in a group. The team found neurons clustered into small world networks with less than two degrees of separation (two neuron hops) and found a clustering rule based on the density of common neighbors. Furthermore, the pyramidal neuron network clusters into multiple groups of a few dozen neurons each that are physically distributed, more than 100 microns apart, allowing clusters to interlace each other. This network pattern was found common across animals.

A more recent paper (Bassett and Bullmore, 2017) reviewed the applicability of the small world network to the brain, based on the most recent studies, and concluded that the concept remains viable in understanding the brain's neural network.

3.14 BRAIN ACTIVITY, SPARSITY, AND NORMALIZATION

When neuroscientists talk of the brain's default mode network (DMN) they are referring to higher-order regions in the brain that show spontaneous activity when an experimental subject is in a state of passivity, and this includes when they are asked to remember events or imagine events (Buckner, 2013). When they are tasked with performing novel non-referential, goal-directed tasks, these areas reduce their activity (Raichle, 2015). According to Talia Brandman et al. (2021), DMN is also active in processing external events: the authors found correlations between DMN activity with states of surprise when subjects were made to watch a movie while under fMRI scanning focused on the DMN.

Research by Matthias Kaschube and his team (Smith et al., 2018) indicates the extent of spontaneous neural activity measured in the brain. Interviewed by Michael Segal (2023), Kaschube finds features in the brain in stark contrast with digital microelectronic circuits which have none: spontaneous activity, correlation, dynamic context generation, unreliable transmission, and noise. The brain also applies context (such as past activity) so that the same identical inputs applied to neurons yield different outputs. Kaschube ascribes much of the noise to neurons that fire action potentials down their axons, but the signals fail to cross over to other neurons. There is also activity going on in parallel in different centers, so the brain can be quite busy, with different centers cross-wired and modulating their outputs. There is also spontaneous brain activity even when the senses have little input, such as in sleep: the brain is as active in sleep as in the wakeful state. Kaschube and his team found in their research that patterns of spontaneous neuronal activity are highly structured in early-stage development of vision in the brain, before structured sensory input exists: this spontaneous activity predicts the local structure of visually evoked orientation columns.

Shy Shoham et al. (2006) find that the brain is sparsely active based on a variety of recording methods, and perhaps contrary to expectations. The authors find many neurons that rarely fire, and they name them "dark neurons". For example, the authors find that less than 20% of pyramidal neurons are detected by recordings. They postulate that neuronal silence may well be the norm in the neocortex. Given the total cost of neocortical energy and the energy cost of a single spike, it yields a neocortical baseline rate of 0.16 spikes per second on average, whereas highly active neurons have been reported to fire 10s of spikes per second. This bounds the number of highly active neurons to 1 in 100 neurons.

Research by Matteo Carandini and David Heeger (2012) shows that the brain performs normalization where the responses from neurons are divided by a common factor based on the summed activity of a pool of neurons. Evidence is accumulating for normalization occurring across many brain regions and across species. Given this widespread occurrence, the authors suggest that normalization is not a feature of a canonical neural circuit but rather a feature of canonical neural computation.

3.15 NEURON MIXED SELECTIVITY

Neuroscience experiments provide evidence that neurons do not activate to respond in a specific functional way but instead can multi-task, responding to inputs combined in different, nonlinear ways depending on context. Fusi et al. (2016) call this behavior "nonlinear mixed selectivity" and as the authors suggest, may have implications for cognitive computation. It is theorized that restricting neurons to perform in a specific way or with only additive, linear mixing (called pure selectivity) restricts the brain's capacity to compute, whereas nonlinear mixed selectivity opens the way for processing complex thoughts and actions. Dimensionality in this context is the minimal number of coordinate axes needed to specify the positions of all points in the neuron output space.

Fusi et al. demonstrate with some simple artificial neural network models that neurons with mixed selectivity can generalize in high dimensional tasks, whereas pure selectivity neurons cannot. To solve high-dimensional tasks, the ensemble must not only possess nonlinear mixed selectivity but also diversity. Diversity is where different mixed selectivity neurons exhibit different response properties, responding to different combinations of values of the task-relevant factors. Fusi et al. point out that for some tasks it may be desirable to reduce the dimensionality of the task. Noise can also impact the ability of ensembles to separate points in the neuron output space, but oscillations may improve this challenge (see Section 9.4).

The concept of mixed selectivity fits in with Rafael Yuste's (2015) suggestion that the neuron doctrine (i.e., that the neuron is the structural and functional unit of the nervous system) may need to be updated to consider ensemble of neurons as physiological units that generate emergent functional properties and states. Yuste goes on to list the types of emergent behaviors that modeling single neurons cannot capture: neuronal reverberations, neuronal assemblies (coalition of large numbers of neurons), neuronal ensembles (modular functional units comprising multiple co-activated neurons that form a circuit), central pattern generators (neuronal circuits producing rhythmic motor patterns for movements such as walking and breathing), and ensemble oscillations.

There are numerous research papers with experimental evidence for mixed selectivity, for example, in primary sensory, decision making, and motor brain areas (Johnston et al., 2020), and Kira et al. (2023) show mixed selectivity neurons emerged through navigation task learning, offering flexibility in decision making made possible by neurons that mix visual and memory information.

Matthew Kaufman et al. (2022) point out, determining whether the mixing taking place is linear or nonlinear or transitioning from one to the other is challenging. However, Johnston et al. showed through modeling that nonlinear mixed selectivity can make orders of magnitude fewer decoding errors than pure selectivity even when pure and mixed forms of selectivity use the same number of spikes and that this benefit holds for sensory, motor, and more abstract, cognitive representations. The authors suggest that nonlinear mixed selectivity may be a general coding scheme exploited by the brain for reliable and efficient neural computation.

3.16 NEURAL OSCILLATIONS

Detecting the electrical signal emanating from an individual neuron or immediate local circuit is too faint unless recorded directly, which is clearly only amenable through intrusive methods. However, it is possible to detect the collective (summed) signal of thousands of neurons through non-intrusive methods such as EEG, which can measure these signals through electrodes placed on the scalp. EEG measures the superposition of currents produced by pyramidal neurons in the outer layers of the cerebral cortex, and the intracellular membrane potentials of these neurons fluctuate, causing a synchronization of signaling at particular frequencies – and this leads to the phenomenon of neural oscillations (Rossi, 2021). While EEG measures electrical activity, magnetoencephalography (MEG) measures the magnetic field produced by the electrical signals and can also be used to observe these oscillations.

Gyorgy Buzsáki and Andreas Draguhn (2004) review neural oscillations in the brain and note that even individual neurons have complex dynamics that lead to resonance and oscillation at multiple frequencies. Neuron assemblies can form synchronized oscillatory patterns and connect individual neurons with behavior. The phase of assembly oscillations, influenced by sensory inputs and internal dynamics of neural networks, carries information that can be used in computation.

A review of cortical traveling waves reveals traveling waves of neural activity in the brain (Muller et al., 2018), occurring through responses to external stimuli or spontaneously generated by recurrent circuits and at two scales: mesoscopic, i.e., single-area and macroscopic, i.e., whole-brain. Lyle Muller et al. (2018) find that

> synchrony of neural oscillations in different neural populations may shape input gain and aid information transfer between these populations, because the synchronous occurrence of action potentials in many neurons is known to increase their ability to drive spikes in shared target neurons. Indeed, experimental evidence suggests that oscillation synchrony modulates the effective connectivity of populations in the cortex, shapes plasticity between neurons and drives processing of stimulus features between visual cortex areas V1 and V4.

Macroscopic waves have propagation speeds from 1 to 10 m/s, in the conductance speed range of myelinated axons. Mesoscopic waves have propagation speeds from 0.1 to 0.8 m/s, in the conductance speed range of unmyelinated axons, e.g., found in the superficial layers of the cortex (Muller et al., 2018).

Oscillations can occur by central pattern generators, Eve Marder and Dirk Bucher (2001) review the role of these generators in controlling rhythmic movements. They point out that contrary to many biology textbooks, rhythmic body movements are not created by reflex activation but by central circuits creating oscillations and these central circuits are not activated by some extrinsic timing information but are activated by neuromodulators. For example, the extension of leg muscles in walking is produced by rhythmic central circuits in which the antagonistic muscles are driven by neurons that inhibit each other. The patterns generated by central circuits are shown in Figure 3.6 and the following text is from Marder and Bucher accompanying their figure:

> Cellular mechanisms underlying pattern generation. (a) Neurons have different intrinsic properties. Some neurons fire bursts of action potentials endogenously (panel 1). In some neurons depolarizing current pulses trigger plateau potentials that outlast the duration of the depolarization but that can be terminated by hyperpolarizing current

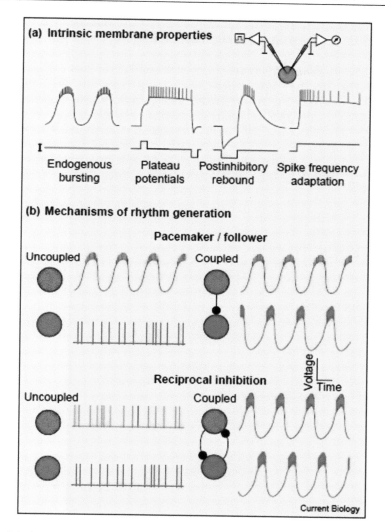

Figure 3.6 Central pattern generation.

Source: Marder and Bucher (2001). With permission.

pulses (panel 2). Some neurons respond to inhibition with rebound firing (panel 3), and others show spike frequency adaptation (panel 4). (b) Rhythms can be generated by endogenous bursters, or can be an emergent property of synaptic coupling between non-bursting neurons. In pacemaker driven networks a pacemaker neuron or neuron (red) can synaptically drive an antagonist (green) to fire in alternation. The simplest example of a network oscillator is one formed between two neurons that fire non-rhythmically in isolation, but fire in alternating bursts as a consequence of reciprocal inhibition.

Central pattern generation is an example of oscillations that occur between two and more neurons interacting.

Research by Klaus Stiefel and Bard Ermentrout (2016) shows that a single neuron firing a sequence of spikes can act as an oscillator. Any process with a regular occurring beat is mathematically an oscillator, and treating a single neuron as well as a network in these terms allows the authors to explore new properties. This treatment allows neuron behavior to be analyzed in phase space: tracking the voltage polarization states as events in phase space. A single-neuron oscillation can be analyzed or the net oscillation of a population of neurons can equally well be analyzed in this way.

Chapter 4

Cognitive processes

4.1 INTRODUCTION

According to Jean Piaget: "intelligence is what you use when you don't know what to do: when neither innateness nor learning has prepared you for the particular situation". The challenge of understanding brain cognition, as image information enters the eyes and then…well, what exactly happens? We know there is a flurry of electrical spike signals traveling between neurons and this translates into a visual experience of the external world inside our heads. This experience is tied up with awareness and consciousness: these attributes of the brain allow for adaptability and choice, decision making in the face of uncertainty. This chapter looks at the neuroscience understanding of cognitive processes.

4.2 COGNITION

Gyorgy Buzsáki (2019) in his book *The Brain From Inside Out*, makes the distinction between thinking of the brain "outside in", which characterizes the brain in terms of external sensory inputs and stimuli, and how the brain evaluates these influences and then reacts to them in some way, in contrast with his preferred view of "inside out" (see also Marcus Raichle, 2010, for a review of the concept, which he terms reflexive vs intrinsic). One of the problems with the outside-in view is that it cannot explain why the same stimuli result in different brain patterns being observed – the brain's internal state is being ignored; this view treats the brain as a passive device. The inside-out view views the brain as an active device with multiple channels of sensory data coming in and evaluated alongside the internal actions of the brain. How the brain reacts to any one sensory input is now influenced by many factors, including the other senses and not least the internal state of the brain.

Buzsáki argues that neurons responding to other neurons firing don't "know" what is causing upstream firing patterns, whether such activations are externally sourced or due to internal causes. The outside-in approach

DOI: 10.1201/9781003507864-6

lacks a grounding model, a means of understanding observations. The inside-out approach lets neurons fire motor actions to say, walk around an object and sense it in multiple ways. This allows neurons to make comparisons and build an internal model of the external environment.

All artificial neural networks to date, including the latest deep learning neural networks (DLNNs), are firmly in the outside-in framework – they lack any internal cognition whatsoever, one merely presses a button to start, ingest some input, produce some output, and that is it, there is no rich hinterland of thought processes, the machine is now "brain dead" until the next input. In contrast we have seen in Section 3.14 how active the brain is when there are no external stimuli, even in sleep.

The internalization of experience, such as thinking in the present, recalling the past, and projecting into the future, are thought processes requiring an active brain, an inside-out brain. Self-organization in the brain can lead to active internal neural exchanges that are not driven by external stimuli (Buzsáki & Moser, 2013; Buzsáki, 2019). Neurons can simulate actions by appropriately firing relevant neurons associated with those real-world actions (e.g., moving a leg, turning the head) but stopping short of carrying out those actions, so that these "actions" are purely internal experiences. The brain's disengagement from external stimuli, "acting" in an internalized virtual world it has created, corresponding to the real world as a reference, but able to manipulate it in multiple ways, defines cognition (Buzsáki, 2019).

There is a corroborating observation by Stanislas Dehaene (2014) that spontaneous excitation in the brain dominates the external sensory signals, to quote:

> Anyone who has ever seen an EEG knows this: the two hemispheres constantly generate massive high-frequency electrical waves, whether the person is awake or asleep. This spontaneous excitation is so intense that it dominates the landscape of brain activity. By comparison, the activation evoked by an external stimulus is barely detectable, and much averaging is needed before it can be observed.

Dehaene estimates that stimulus-evoked activity accounts for probably less than 5% of the energy consumed by the brain.

The inside out view of the brain is essential to account for this activity, where the brain is an autonomous "machine" generating its own thought patterns in ceaseless activity.

4.3 HUMAN CONSCIOUSNESS

Consciousness was at one time a taboo subject for research avoided by neuroscientists. One of the current pioneers in the subject, Dehaene (2014) recalls how this began to change in the late 1980s to the point where it is now a major focus of research. Scientists have a habit of only recognizing a topic if they can grasp it in some way, and consciousness research benefited from new techniques (see Appendix) that provided an entry to understanding it.

Dehaene (2014) offers a definition of consciousness, made up of three components. First is vigilance or wakefulness, as opposed to being asleep. Second is attention which is information that enters the brain but not necessarily at an awareness level. Third is conscious access – this is the current awareness or focus of the mind. To summarize, to be conscious Dehaene requires three properties to be simultaneously active or true:

1 Awake
2 Attentive
3 Aware

Neuroscientists have been investigating the neural correlates of consciousness, connecting mental states with activated neurons (Tononi & Koch, 2008). In recent years this has been refined as neural correlates of consciousness and neural correlates of attention: there is a distinctive difference between consciousness and attention (Nani et al., 2019). According to Andrea Nani and team, the function of consciousness is to create "a continuous and coherent picture of reality", while "attention has the function of attributing relevance to the objects of thought".

In blindsight experiments by Dehaene and team they show the brain paying attention to sensory information on the fly and processing it without this information necessarily rising into consciousness. These experiments were part of a program to identify a signature of a conscious thought: a neural firing pattern taking place within milliseconds of a change in sensory input. Experimental subjects were shown images of white masking noise and then shown a group of numbers and asked to estimate the average of the numbers. Unknown to the subjects, numbers were interspersed in the white noise for a microsecond, too quick for the brain to be aware but the signal was recorded by the brain (shown up in brain imaging). When these subliminal numbers were low the average estimate tended to be low and when they were high the estimated average was high, demonstrating the difference between attention (operating in the subconscious) and awareness. Further experiments by Dehaene and team recorded a brain imaging signature for detecting conscious awareness of new information.

Gerald Edelman and Giulio Tononi (2000) point out that in the conscious state, we can only hold up about seven separate items of information, and decisions can only be made serially within a few hundred milliseconds of each other – this duration is called the psychological refractory period (Pashler, 1994). Moreover, consciousness is a series of states that appear to us as seamless but are hops of around 100 ms apart. Edelman and Tononi postulate that many neurons must interact rapidly and reciprocally through a process they call re-entry – if re-entry is blocked then consciousness disappears or shrinks.

The reciprocity of neural connections is well known, for example:

> The first principle is that of reciprocity of corticocortical connections. More than a decade ago, it was noted that pathways within the visual cortex tend to be bidirectional, such that if area A projects to area B, then area B is likely to project in turn to area A.
>
> (Felleman & Essen, 1991)

Edelman and Tononi note that activity of neuron groups supporting consciousness experience continually changing patterns and if, contrarily, many neurons fire in a repeated way, this results in unconscious states such as deep sleep and epilepsy.

A study by Du et al. (2023) on the global organization of the cerebral cortex reveals that it can be thought of as a three-level network hierarchy:

- *Level 1 (low level)*: locally organized sensory and motor networks.
- *Level 2*: networks that link to distant regions.
- *Level 3 (high level) and beyond*: networks that populate and connect widely distributed zones of higher association cortex. Within level 3 there are distinct network clusters that link to networks that have different task responses: language, social, and episodic functions.

As Dehaene describes it, early visual neurons initially process a limited window of the retinal input processing it in relative isolation, without any awareness of the overall picture, but higher association areas of the cortex break the nearest-neighbor modularity of early cognitive operations and connect to major connection hubs. Dehaene notes that the cortex is full of loops and bidirectional projections (the neural reciprocity noted above). Neural network simulations by Dehaene show that this type of connectivity leads to self-organization and attractor states, meaning stable and reproducible activity patterns with long duration.

The back-and-forth communication between near neighbor neurons is Gerald Edelman's re-entry communication mentioned above (see also Edelman & Gally, 2013), but Edelman also notes that re-entry includes links between similar parts of the two hemispheres of the brain, and that re-entry connections are not merely one-time feedback but part of a continual recursive process.

We should also note the sparsity of the active neurons, at any given time most neurons are not firing: conscious content is characterized by many silent neurons and a few active ones. Part of this has to do with how those neurons that are active are also activating inhibitory neurons that silence large parts of the brain, at least for the duration of the "thought" over a few hundreds of milliseconds, until the next wave of activation patterns. According to Shoham et al. (2006), up to 90% of neurons are not active at any given moment, with some neurons firing rarely or only to specific stimuli. Shoham and team say that while "interneurons and cerebellar Purkinje cells are active most or all of the time, the diversity of cases in which many neurons appear to be silent includes major neuron types in the mammalian neocortex and hippocampus, and the cerebellum".

Consciousness appears at the higher levels of the above model, at areas of the brain that have made sense of the information percolating up from the lower levels, the hierarchical structure is essential to shield consciousness from too granular a level of information that would only confuse a person, awareness of the sub-conscious would not be useful.

4.4 ANIMAL CONSCIOUSNESS

Neuroscientists (e.g., Seth, 2021) increasingly acknowledge that animal consciousness exists, although most likely different from ours. Research providing evidence for consciousness in animals is at an early stage of development and has moved from a dismissive "animals have no consciousness" to "how do we measure it". Research shows that higher animals such as apes, elephants, octopuses, and dolphins have consciousness – see, for example, the Cambridge Declaration on Consciousness (2012) which describes a consensus among scientists that all mammals and birds, and many other creatures possess some degree of consciousness. Jonathan Birch et al. (2020) argue that different animals have consciousness that varies in degree on a sliding scale from low to high abilities but also this scale needs to be multi-dimensional to reflect different attributes of intelligence. They offer the following dimensions for investigation:

1 *Perception – visual*: how rich is the animal's visual cognition.
2 *Perception – touch*: how rich is the animal's cognition based on touch.
3 *Evaluative richness*: the spectrum of emotions from pain, fear, grief, anxiety, feeling bad to pleasure, joy, comfort, love, and feeling good.
4 *Unity of perspective*: a unified view of self and environment integrating all senses at a point in time.
5 *Sense of time*: integrating experiences across time to yield a sense of time direction.

To which I believe should be added:

6 *Other sensory*-related cognition: smell and sound is important for animals with poor or non-existent sight, and electricity for some aquatic animals.
7 *Sense of self-awareness* at the highest level of consciousness.

Some tame animals kept in captivity show great intelligence, e.g., Washoe the chimpanzee (Fouts et al., 1989) was taught sign-language, and clever parrots solve cognitive tasks (Pepperberg, 2006). In birds, the neurons are smaller than in mammals and Suzana Herculano-Houzel (2020) notes in the bird pallium (particularly corvids), neuron numbers are 0.5 to 2 billion, containing more information processing than in equivalent-sized mammalian cortices. The bird pallium is the mantle in the brain that corresponds to the cerebral cortex in mammals. Herculano-Houzel therefore contends that birds possess consciousness. A good overview of research into avian intelligence is given by Nathan Emery (2016), revealing that birds have precise memory of past events, can plan, and have versatile tool use, amongst other advanced cognitive behavior.

Barron and Klein (2016) argue that a marker for consciousness is subjective experience, and subjective experience in turn is supported by structures in the midbrain that simulate the movement and location of the animal in its external environment. These midbrain structures are present in vertebrates and the authors argue that insects in the invertebrate class possess somewhat analogous structures, and therefore by extension, insects possess some form of consciousness.

After extensive observations and experiments, Lars Chittka (2017) finds sufficiently complex behaviors in bees to consider them conscious creatures, behaviors not normally associated with their natural lives. For example, honeybees can count three items. In another experiment, a bumblebee must pull a string to bring a container with nectar from under a cover and be able to drink it. A bee can be taught to move a small ball and receive a reward when positioned in a certain spot, while other bees observing this experiment learn this skill and are similarly rewarded. In these and other experiments Chittka argues a bee understands the outcome of its actions and that of other bees, and this requires a degree of consciousness.

Theory of mind (ToM) is the name cognitive scientists give to the ability to ascribe mental states, such as desires and beliefs, to others. In a review of ToM in animals by Krupenye and Call (2019), the authors conclude:

> Research has already made huge strides in identifying ToM capacities in nonhuman animals from corvids to primates to dogs. Recent advances in studying cognitive mechanisms have further blurred the lines between humans and nonhumans, raising the possibility that some of the richest ToM abilities, such as understanding of subjective desires and false beliefs, may not be the exclusive province of our species.

Finally, to a phenomenon shared by animals and plants: anesthesia. Plants and animals shared a multi-cellular evolutionary history before diverging some 1.6 billion years ago. Sonke Scherzer et al. (2022) noted that while the Venus flytrap and plants in general do not possess neurons, the flytrap fires action potentials (hapto-electric signaling) in response to its trigger hairs touching an object (ideally a fly from the plant's perspective) and closing the trap. Scherzer et al. applied ether, which has an anesthetic effect on human and other animal brains, to the Venus flytrap, and found that it prevented the trapping mechanism from operating.

4.5 SENSORY THINKING

This chapter focuses only on the visual capabilities of the brain as the foundation for thinking in humans. Early humans acquired these capabilities before language and therefore visual consciousness is a foundation for intelligence. Thus, for the first steps in building HLAI I'm being selective and ignoring language for now. There is also the consciousness associated with the other senses, but again to keep the material focused I'm also leaving aside other sensory inputs to the brain, although I consider these also to be more important (because foundational) than language for our purposes.

Temple Grandin (2022) has written about her experiences as a visual thinker. She describes visual thinking as not being about vision but rather how the brain processes information in a visual mode of thinking versus the more usual verbal, internal dialogue mode of thinking. Furthermore, current research finds two modes of visual thinking: one is a spatial visualizer that thinks in patterns and abstractions, and the other is an object, pictorial visualizer (Kozhevnikov, 2013), which is how Grandin describes herself. When a person speaks to her, she instantly translates the words into pictures in her mind. She also describes how she thinks by running visual simulations in full 3D in her imagination. She can recall details of her environment in her head and repeat image sequences over and over and investigate design changes. Her mind if left free thinking will jump by associations to other images, and to think in a focused way she just brings her mind back to the original focus of attention.

Most people ("normal" or neurotypical) think verbally, whereas a small minority think visually. As Temple Grandin points out there is a spectrum of types of human brain with purely verbal and purely visual thinking at each extreme. A difference between verbal thinkers and visual thinkers is that the former think sequentially and like order, whereas the latter see images in their mind, are good at seeing associations, and like maps. Grandin can recognize a location after only visiting it once before. Visual thinkers are good at working out how a mechanical device works. A visual thinker is also super alert to their environment. A great ability with jigsaw puzzles is an indicator of visual thinking, as is memorizing large amounts of information. Mnemonists use an ancient Greek technique called the memory palace where items to remember are associated in vivid ways with the different rooms of an imagined palace. Visual thinkers can naturally perform such memory associations. Peper (2020) points out advantages of visual thinking for the individual is precision, richness in detail, ability to imagine in 2D and 3D, and deal with complex scenarios.

Research into visual thinking shows that children typically start out as visual thinkers (experimental subjects at five years old) and then switch to verbal thinking at a later age (experimental subjects at 11 years old), see Koppenol-Gonzalez et al. (2018).

In the evolution of humans there was a point before hominins had language and therefore these ancestors of ours were visual thinkers, as vision was a mature evolutionary feature well before the brain attained language. This implies that visual thinking was the sole and normal mode of thinking in early humans and perhaps explains how young children today start out as visual thinkers before acquiring language skills. Visual thinking, exploiting the existing vision system, would be the first mode of intelligent thinking. In humans the processing of vision takes up a third of the brain and is responsible not just for external visual processing but internal visions such as dreaming.

Cataldo et al. (2018) performed experiments that reflected the efficacy of just vision and gestures versus speech alone in stone toolmaking. Furthermore, this work indicates that early hominins (before the evolution of language) were successfully using gestures, vision, and significantly, visual thinking, to build tools and perform cognitive tasks.

4.6 VISION AND THE RULES OF PERCEPTION: VISUAL INTELLIGENCE

Vision is not only the dominant sense in many species but also takes up the largest proportion of the brain (Ptito et al., 2021). According to research by Sheth and Young (2016) vision processing occupies 20%–30% of the human brain.

Thus a considerable amount of brain resources is dedicated to processing the 2D visual signals from our eyes and its function is to create the external world in our head: objects, colors, textures, depth, shape, and motion. We do more than just register this external reality – we understand what it is: this is visual intelligence. Visual intelligence as a label has several uses, but our interest is in the neuroscience study of vision in the brain. Donald Hoffman (1998) talks of visual intelligence as an intelligence that we take for granted, not aware of its presence because of how well it is embedded in our thinking.

A human baby learns how to see in the first year of life. Vision presents a challenge to the brain: the external world is in 3D, but our eyes present us with 2D images and there's an infinity of possible 3D interpretations. The brain resolves this ambiguity, of seeing the correct interpretation, through having innate rules, Hoffman calls these *rules of universal vision*, and with these rules, each baby goes through a subconscious learning process to make sense of visual information. At the end of this period, the baby has acquired what Hoffman calls *rules of visual processing*, allowing the baby to construct visual scenes in the head through looking at the external world. The first set of rules are provided by biology and the second are learned and may be culturally influenced. The visual process is multi-staged, starting with neurons that recognize light and dark boundaries, patterns of lines and shapes, and as these neurons process from one stage to the next, an image of a complete object is built up at higher levels of the chain of neurons. Hoffman (1998) devotes his book to explaining the various *rules of visual processing*, for example, this is rule 5 chosen at random: "Always interpret a curve that is smooth in an image as smooth in 3D".

Vision is not a passive activity, it is an active process of construction, all performed unconsciously. Furthermore, the visual process does not always rely purely on visual information but can also use any of the other senses, touch, sound, smell, and taste, to help form the image construction.

Returning to the challenge of how the brain turns its 2D input into a 3D image – it adds depth to its 2D construction. Depth construction becomes a habitual process, but it also follows certain rules, and this is also a source of weakness: the rules exclude possible constructions that don't conform to the rules. While their value is in reducing infinite possibilities to one or two views, they can also lead to paradoxes and illusions. For example, in the case of two valid images, the brain will flip between the two views, and this is the source of many optical illusions – the artist Escher is famous for his "impossible" drawings.

To understand vision, what we perceive and how we respond to it, we need to discern between what Hoffman calls the *phenomenal* sense of vision and the *relational* sense of vision. Hoffman provides a useful metaphor to explain the difference: if you consider how you interact with a software application on your computer, say a document editor, the phenomenal sense is the construction of the software code, the computer programming that created the application. The relational sense is how you interact with the application when it runs. The application user does not need to know anything about computer programming and how the application is built to use it, a different set of skills are required to be a proficient user of the document editor. Returning to the brain, the construction of how we perceive is the phenomenal sense, it provides the apparatus for vision and supports our sense of vision, which is the relational sense, this is how the brain thinks about that visual information. The weaknesses of the phenomenal sense, the gaps in the construction rules, lead to paradoxes in the relational sense.

The human eye contains about 100 million neurons and image processing performed by the retina creates dots of sensory information. Hoffman makes the point that as the retina images are dots, then the brain must turn this information into a construction of smooth lines, curves, and surfaces.

Everything that we see is a construction in our brain: from colors to motion. Hoffman describes a stroke victim whose ability to see motion disappeared (akinetopsia: motion blindness); in all other respects, her brain behaved normally (she was able to perceive motion through sound or touch). This reinforces our understanding of the extent to which our view of the world is a construct in our head from basic sensations.

Visual intelligence is connected to our emotional intelligence and our rational intelligence. When the connections between these intelligences are impaired in any way, through strokes or accidents that damage the brain and these connections, a disconnect appears in our understanding of the world, illustrating the importance of how the different intelligences combine to provide a holistic and accurate view.

Experiments reveal that the brain holds an internal model of the external environment, for example, the ability to interact with objects in our immediate vicinity that are obscured to our vision, and moreover, this model remains consistent as we move in the environment so that obscured objects are re-positioned consistently with our moving location and its location in the brain is strongly linked to our motor system (Land, 2014).

Chapter 5

Time and space in the brain

There are two questions that arise when considering time in the brain: first, how time affects the processing in the brain and second, how the brain creates the experience of time in our consciousness, i.e., how do we make sense of thinking about the past, present, and future.

Starting with the first question, most animals are impacted by the day and night cycle of the Earth's daily rotation about its axis (for exceptions, see Bloch et al., 2013). This circadian rhythm affects the human brain, the question as always in this work, is whether and how it impacts intelligence. It clearly impacts the human body: circadian rhythms control our sleep cycle, blood pressure, metabolic rate, and more.

The body keeps an internal circadian clock which research shows to approximate a 24-hour cycle, but not exactly, so a coordinating correction mechanism is necessary, called entrainment in circadian biology (Foster, 2022). Of the 86 billion neurons in the human brain, about 50,000 neurons in a region called suprachiasmatic nuclei (SCN) in the hypothalamus perform a combined role of "master biological clock", each one keeping its own cell-level molecular clock and connected to its neighbor. SCN sits near where optic nerves enter the brain, including some that enter the SCN, providing light/dark information from the eyes.

A Nobel prize was awarded for the discovery of how the SCN molecular clock worked (Hall et al., 2017): it operates through a negative feedback cycle, as follows. Clock genes in the SCN neurons guide the creation of clock proteins, formed in the cytoplasm surrounding the nucleus. The clock proteins interact and form a protein complex which moves into the nucleus and inhibits further production of clock proteins. Over time, the protein complex is broken down, which allows the nucleus to start a new cycle of clock protein creation. One cycle takes about 24 hours, and the clock neurons generate a periodic signal – Russell Foster (2022) says electrical or hormonal, a continuing research area – that then spreads throughout the body, coordinating a circadian clock. Jeffrey Hall, Michael Rosbash and Michael Young performed their research on the fruit fly *Drosophila melanogaster*, with which we share the same clock biology (and a common ancestor around 570 million years ago), illustrating how evolution reuses its

DOI: 10.1201/9781003507864-7

inventions over again, and how connected humans are with all life forms on Earth, including insects.

Scientists have since discovered that there exist peripheral clocks in various body tissues, driving a circadian rhythm in their respective physiological functions (Richards and Gumz, 2012), all using the same negative-feedback protein cycle. These peripheral neurons (which run to billions) keep time with contact by the SCN clock neurons and without that connection become uncoupled and lose time synchrony. The SCN clock neurons align the body clock neurons to the external day and night cycle.

Foster shows that the internal cycle of human growth hormone peaks by a factor of three for a period of about two hours with the onset of sleep, which affects neural development and brain functions such as learning and memory consolidation. During sleep new memories are formed, with the hippocampus playing a central role (Bendor and Wilson, 2012). The hippocampus contains time cells that track when an episodic memory occurred (Umbach et al., 2020) and place cells that track where the memory occurred. There occur spontaneous replays of recent experiences in the hippocampus and stabilizing their memory in the neocortex, following which retrieval from the neocortex is independent of the hippocampus. Sleeping has been shown to be important for consolidation of memory (Schönauer et al., 2013), and other cognitive tasks, reinforcing anecdotal evidence of people solving problems and gaining insights on waking up after a night's sleep (I believe the trick is to recite the challenge just before shuteye).

Answering the second question, how the brain creates the experience of time in our consciousness, has posed more of a challenge to neuroscientists. Gyorgy Buzsáki (2019) provides an overview of current progress in his book. Space and time (in the classical Newtonian world of our experience) are intertwined; for example, your watch measures the passage of time by the spatial position of the watch hands. In the brain, distance and duration are believed to be represented by the same structures and likely by the same neurons.

Experiments on rodents show neurons that have learned to fire when the animal visits a specific place and when positioned at a grid point in a virtual map, with one neuron linked to a unique place, and one neuron linked to a grid point (Weber and Sprekeler, 2018). Together, these place and grid neurons generate a virtual 3D Euclidean space map identifying points of interest to the animal and relative locations (Buzsáki and Llinas, 2017). Buzsáki and Llinas point out that hippocampal circuits are blind to the modality of cortical inputs: messages are processed in the same way regardless of their origins: space, time, sound frequency, odor, memory, or something else. Buzsáki and Tingley (2018) go on to suggest that the hippocampus operations respond to the internal process of sequential activity and rate of change of neuron firing, and this is sufficient to process time and space in the brain.

Marc Howard (2018) has proposed that the brain uses a common computational language for visual attention and memory retrieval. Given the greater granularity of detail in the center of vision with decreasing granularity for signals received in the periphery of vision, Howard proposes that time functions and memory are processed in a similar way with the present receiving the greatest attention and the past and future having decreasing attention the further they are from the present and presents evidence for a logarithmic compressed representation of time.

Our processing of events is granular, for example, vision processing in the brain integrates all incoming signals in a time window of around 100 ms, anything that changes faster than that is perceived as a single image (of course this is how cinema works – we don't see a movie as a series of static images, which they are, but rather as a moving picture). A complication arises when visual, sound, and other sensory signals associated with an event are processed in the brain at different speeds (due to different brain circuits being involved) and must be combined to form a perception of a single event. The brain solves this problem by waiting about a tenth of a second, resolving all the incoming signals into a consistent event (Eagleman et al., 2005). The drawback is that we perceive events in the past (albeit only 0.1 second lag). Eagleman and team showed that the brain calibrates incoming events to make holistic sense of the disparate signals arriving in the brain.

Part Two

Theories, models, and algorithms

INTRODUCTION

This part provides a survey of theories, models, and algorithms mostly in AI research, but we include theories of consciousness as well. The emphasis is on what is sufficient in modeling the human brain to capture the essential behavior that results in intelligence. This is where computational neuroscience parts ways with AI researchers. Computational neuroscience uses mathematical models, is highly empirical and phenomenological, and is evidence based, and has found application in helping build interfaces for connecting prosthetics to humans with disabled limbs. However, these models have not cracked the understanding of the neural code and I have not found a computational neuroscience model or theory that addresses the mission of this book.

I have not reviewed neuro-symbolic AI approaches in this survey, despite the recent resurge of interest in the topic (Garcez and Lamb, 2020; Amit et al., 2023). Neuro-symbolic AI aims to combine neural network methods such as deep learning with symbolic reasoning to bring out the best of both approaches. Most such work is pre-occupied with natural language processing (NLP), and this book has avoided delving into NLP for reasons discussed in Part Three.

DOI: 10.1201/9781003507864-8

Chapter 6

Theories of consciousness

A view of what is mind is given by Stan Franklin in his book *Artificial Minds* (1999) and depicted in Figure 6.1. In this view of mind, which shares the "society of mind" concept with Minsky (1988) and others, the mind comprises autonomous agents that may communicate with each other, but which operate independently of each other, i.e., there is no top-down hierarchical command and control for all the activities of the brain. Rather, these agents either perform unique functions unhindered or compete or cooperate with each other for the "right to act" in serving common goals. These agents may be "mindless" but the sum of their activities creates an intelligent mind. This view contrasts with Allen Newell's (1990) program of creating a unified theory of cognition, discussed in Section 8.2.

In Franklin's view of mind (in the animal kingdom), it is a continuous spectrum of states, of degrees of mind. As expressed above, the mind is the aggregate action of multiple autonomous agents: they only communicate with each other when necessary, and this view is counter to unified theories of cognition which assume a command-and-control architecture. The mind's activity is to perform "next action" which will occur in parallel across the agents and which is continuously performed rather than discrete. The mind *produces* information from sensory inputs, it does not extract information from the environment, rather it creates information – this points to information as being subjective rather than objective and existing externally: different minds processing the same sensory inputs can produce different information. Finally, for Franklin, the mind uses memories of past information and experiences to produce new actions, and these memories are reconstructed from pulling together salient cues that are stored.

I briefly mention three theories of consciousness, in-depth analysis is outside the scope of this book, but the references below and Anil Seth (2021) are recommended reading:

- *Neural Correlates of Consciousness* (NCC): NCC was proposed by Christof Koch (2004) to connect mental states to physical neurons and their states. The theory proposes that conscious states directly correlate to a minimal set of neural activities. This theory was also

                                              DOI: 10.1201/9781003507864-9

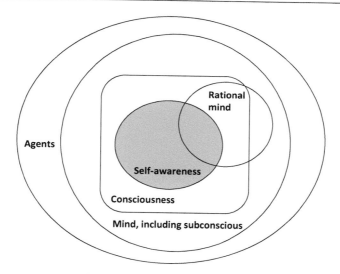

Figure 6.1 A view of mind.

Source: Based on Stan Franklin (1999).

championed by Francis Crick, who wrote that nerve cells and molecules give rise to consciousness.

- *Integrated information theory* (IIT): Created by Tononi (2004), this is a mathematical theory that treats information as a property of the universe, like mass or energy, and that consciousness is similarly a property of the universe, measurable by a single quality, called "phi" in the theory. IIT has many critics, for example, it is considered panpsychism, where mind/consciousness is classed as a feature of nature, and most recently, it has been called out as pseudoscience by Stephen Fleming et al. (2023).
- *Global workspace theory* (GWT): This theory is championed by Bernard Baars (2021), Stan Franklin (1999), and Stanislas Dehaene (2014). GWT involves continual interactions between subconscious and conscious states. The global workspace is a memory where thoughts turn conscious and last a few seconds at a time and is analogous to a theater stage where the spotlight focuses the attention of the conscious mind. Dehaene's work on consciousness, covered in Section 4.3, is inspired by GWT.

A note on panpsychism: it is a philosophy of mind that had many adherents, including Plato, Spinoza, Leibniz, and Bertrand Russell and continues to attract supporters. The paradox in panpsychism is that human consciousness exists because we humans experience it, but that is not necessarily evidence for consciousness as some kind of irreducible feature of nature.

Chapter 7

Neurorobotics
Embodied AI

This chapter is not an exhaustive historical record of research in the field of building robots with AI that move and interact in the real world. Rather, it is a sample of current ideas in this field. Note, the term "embodied" is often used to refer to a physical AI machine embedded in the real-world environment. An embodied AI machine or robot is also referred to in the literature as a neurorobot. Ziemke (2003) identifies six different notions of embodiment as there are contrasting views on what kind of physical body (if any, i.e., it could be simulated) is required for embodied cognition:

- *Structural coupling* between agent and environment (does not require a body), i.e., each one can perturb the other through connecting channels.
- *Historical embodiment* as the result of a history of structural coupling, i.e., the connection between the agent and environment may have been in the past and affects the behavior of the agent in the present environment.
- *Physical embodiment*, this requires a physical instantiation of the agent beyond virtual/software.
- *Organism-like embodiment*, i.e., has organism-like bodily form such as humanoid robots, both living and artificial agents.
- *Organismic embodiment of autopoietic kind*, i.e., a living system capable of growing or creating its own parts and therefore autonomous, in contrast with machines that are assembled in a factory (allopoietic) and are therefore governed by external forces (heteronomous).
- *Social embodiment* is the ability to communicate through body language.

So, while humans can be regarded as embodied cognizers, there is no consensus on what kind of body an AI machine would need to possess for HLAI.

The motivation for this field is nicely put in the paper by Juyang Weng et al. (2001) when defining autonomous mental development:

 DOI: 10.1201/9781003507864-10

With time, a brain-like natural or an artificial embodied system, under the control of its intrinsic developmental program (coded in the genes or artificially designed) develops mental capabilities through autonomous real-time interactions with its environments (including its own internal environment and components) by using its own sensors and effectors.

In other words, to achieve in AI machines human-level capabilities in vision, speech, movement etc., the AI machine must exist in the real world and interact with it, as part of its learning process.

The application of evolution inspired techniques to breed embodied intelligent robots with HLAI has been pursued for many decades, e.g., Pfeifer and Bongard (2007). Building on the work of Pfeifer and Bongard, Krichmar and Hwu (2022) suggest the following design principles to be used in designing future neurorobots, such that they must

1 React quickly and appropriately to events.
2 Be able to learn and remember over their lifetimes.
3 Weigh options that are crucial for survival.

In explaining these principles, Krichmar and Hwu note that to react instantly to events, decisions and actions must be taken at the periphery of the body, as there is insufficient time to engage the central nervous system and brain. These instant morphological computations are performed at the periphery by engaging sensors, actuators, and reflexes. This is an important principle, that the body periphery is acting without direct guidance from the brain. Where senses, morphology, and environment allow responses to be kept local at the periphery, this speeds up response time and minimizes energy consumption.

Minimizing energy use while maintaining high performance is a general principle applied by the brain and it can achieve this through sparse coding in how neurons fire when communicating with each other (Olshausen and Field, 2004; Beyeler et al., 2019). Furthermore, the small world architecture of network connectivity is believed to apply to brain neural networks (Sporns, 2010), this model is energy efficient and performant.

A recent framework for bringing together all the strands of current AI advances, such as deep learning, with an embodiment in a physical dynamical environment is proposed by Moulin-Frier et al. (2017), called Distributed Adaptive Control for Embodied Artificial Intelligence, and is suggested as the route to achieving general AI.

The current state of embodied AI research is considered by Hughes et al. (2022) and suggests the research community requires a grand challenge to focus research. A survey of nine research simulators for embodied AI is conducted by Duan et al. (2022), looking at features such as whether the environment is games based or world-based, degree of physics real-ism, and more. An example of a simulator is the Neurorobotic Platform (NRP) from EBRAINS, a project funded by the Human Brain Project and the EU. The latest model NRP 4.0 was released in 2023. It can be accessed online here: https://www.ebrains.eu/modelling-simulation-and-computing/simulation/neurorobotics-platform/, and here: https://neurorobotics.net/access-the-nrp.html. EBRAINS describes its users as (I quote)

- *Neuroscientists* who want to test and refine their models of brain functions in closed loop experiments.
- *Roboticists* who strive to deliver on the promises of neuroscience-based embodied Artificial Intelligence.
- *Anyone* who believes that embodiment is the way forward for both brain research and robotics.

Chapter 8

Engineered brain architectures

8.1 INTRODUCTION

Cognitive architectures are top-down constructions that their designers believe are the essential components of thought processes in the human brain. The systems described here are architectures, and their implementations are sometimes called agents, which may add task-specific components that go beyond the parent architecture.

I refer the reader to three recent accounts that provide a historical narrative of this branch of AI for filling in the gaps of the limited coverage here: a recent review of the field by Kotseruba and Tsotsos (2020), an extended framework reviewed by Dodig-Crnkovic (2021), plus Lieto (2021) and Bermudez (2023).

There is an important point to note: researchers in this AI branch are not necessarily looking for biologically realistic or constrained models; those that go beyond such realism cite the analogy of airplanes versus birds: both fly in the air, and both are quite different in many respects.

8.2 COGNITIVE ARCHITECTURES

Part Zero mentioned the two main ideas that fought for dominance (and funding) in AI research for many years: the symbolist approach and artificial neural networks. Symbolism assumes the metaphor of the brain as a computer and its mind as the software running it. To stretch this metaphor further, the proponents of this view sought to discover the "computer language" of the mind, a language of thought. Artificial neural networks research went through re-labeling during funding difficulties, the so-called AI winters, such as connectionism, parallel distributed processing, and most recently, deep learning, except of course that the latter most is now famously successful, and its neural network basis is proudly re-asserted. An essential difference between symbolism and neural networks is in the underlying metaphor: in neural networks the mind *is* the brain – there is no dichotomy between mind and brain (Eliasmith, 2013). We turn to neural

DOI: 10.1201/9781003507864-11

networks in Section 8.6 onwards, but first look at the symbolic and hybrid approaches that combine symbolic and neural networks.

Chris Eliasmith in his book *How to Build a Brain* (2013) sets the foundation for cognitive architectures in the form of core cognitive criteria for theories of cognition, i.e., basic features of human cognition that must be explained by a cognitive model *theory* (note some of these features are language oriented), and which assume that thinking is produced by computational procedures operating on mental representations (Thagard, 2012):

Core cognitive criteria:

1 *Representational structure*
   a *Systematicity*: representations (thoughts) are connected (through context or intention) and sets of representations can be intimately linked.
   b *Compositionality*: the meaning of complex representations can be decomposed into basic representations, i.e., complex representations are the addition of basic representations.
   c *Productivity*: generate many representations from a few basic ones.
   d *Binding*: many basic representations must be bound together to build a complex representation.

2 *Performance concerns*:
   a *Syntactic generalization*: exploiting the structure of language independent of its content.
   b *Robustness*: insensitive to noisy or missing data.
   c *Adaptability*: can learn, updates future performance based on experience.
   d *Memory*: can model a range of different types of memory, e.g., short term and long-term memory.
   e *Scalability*: the model must scale from addressing small challenges to large complex problems.

While a symbolist approach may represent knowledge in terms of explicit rules, a neural network distributes its knowledge implicitly across the network. Incidentally, a symbolist model enjoys the property of transparency, whereas a traditional neural network is like a black box, so understanding how knowledge is represented in such a neural network and why it produces a given result is more difficult to establish (but there are techniques that can help).

John Laird (2022) describes cognitive architecture as a task-independent infrastructure that learns, encodes, and applies an AI agent's knowledge to produce some action. Furthermore, it is a software implementation of a general theory of intelligence.

In this section, I have selected examples among the most popular cognitive architectures: Soar, ACT-R, and Adaptive Resonance Theory (ART). In a review of the field "40 years of cognitive architectures: core cognitive abilities and practical applications" by Iuliia Kotseruba and John Tsotsos (2020), the authors reported some 84 models actively pursued across 900 research projects. In their analysis of the practical applications of cognitive architectures, ACT-R is by far the most popular, followed by ART, Soar, and hierarchical temporal memory (HTM) (which we cover in Section 8.5). Eliasmith's semantic pointer architecture (SPA) comes in at position eight – we cover it in Section 8.2.3. Most of the models reviewed by Kotseruba and Tsotsos when applied to new tasks the accumulated knowledge cannot be reused. In the case of SPA and its use of sparse techniques it can perform up to eight unrelated tasks.

8.2.1 Soar

Building on the work of Allen Newell and his unified theories of cognition (Newell, 1990), Soar was developed as an example of a unified cognitive architecture, supporting multiple theories (micro-theories) that each addressed one aspect of cognition and brought them together in one system. Soar was therefore conceived to solve multiple cognition challenges through computer models (Laird and Rosenbloom, 1993; Laird, 2022). Soar offers computational building blocks for creating AI agents. Soar is free open source software, maintained and available at soar.eecs.umich.edu.

Soar has a symbolic architecture, with a clear separation of the control structure of the program from the operator and knowledge about the task. When Soar attempts to fulfill a goal, it applies several approaches (Laird, 2022): reasoning and retrieval from external sources, reasoning with procedural memory, reasoning with non-symbolic knowledge, retrieval of a solution from episodic memory or semantic memory, or interaction with the external world.

If Soar hits an impasse in trying to fulfill a goal, it applies an automatic operator subgoaling technique that generates a subgoal and pursues that as the current goal. It will continue the subgoaling process until it solves a goal, and then winds back up the tree of subgoals to the original goal, returning the knowledge that allows progress. Depending on the model, Soar can use means-ends analysis to limit search in a problem space, and it does this by selecting an action that minimizes the difference between the current state and the goal state, this minimization continues recursively. Clearly, for this to work, the system must be able to select the right action to produce the minimization and then be able to detect the changed state.

Figure 8.1 shows the key components of Soar. The latest version of Soar has three forms of long-term memories, where, as the system gains experience solving problems, knowledge is accumulated:

- *Procedural*: how and when to do things.
- *Semantic*: facts about the world.
- *Episodic*: memories of situations experienced.

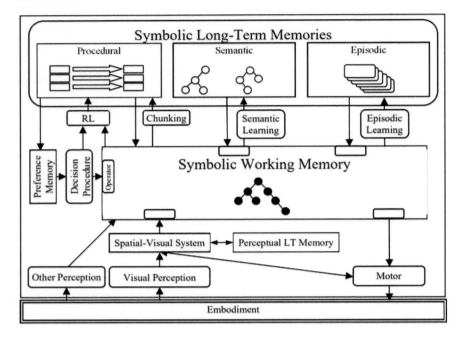

Figure 8.1 Structure of soar memories, processing modules, learning modules, and their connections.

Source: Laird (2022). With permission.

These memories connect with a working memory which contains knowledge specific to the current situation. All tasks are represented as problem spaces, and long-term memory comprises "production systems" that satisfy the production of tasks. Knowledge search involves all productions firing and the best fit operator to the query is selected. The term "chunking" in Figure 8.1 refers to the knowledge gained when a goal impasse has been resolved – this knowledge is called a chunk that creates the production that embodies the knowledge gained. If the same impasse recurs, it will retrieve the knowledge instantly from long-term memory. Chunking converts goal-based problem solving into long-term memory and is an automatic, continuous learning mechanism that applies to whatever Soar experiences (Newell, 1992). The external inputs are either visual or "other perception" which includes language (text or verbal) and hence natural language processing is applied to extract meaning from the input.

Soar has been tested in some large-scale trials; for example, a version called agents TacAir-Soar and RWA-Soar were created for training pilots in large-scale military simulations, and capable of executing most of the airborne missions that the U.S. military flies in fixed-wing aircraft (Jones et al., 1999).

8.2.2 ACT-R

Adaptive Control of Thought-Rational (ACT-R), first developed by John Anderson and Christian Lebiere at Carnegie Mellon University and inspired by the work of Allen Newell, started out as a symbolic cognitive architecture but has evolved into a hybrid model integrating symbolic AI and neural networks (Anderson, 2007). ACT-R is implemented in Common Lisp and has its own coding language to create ACT-R agents. ACT-R has an active maintenance community at act-r.psy.cmu.edu and makes available model libraries. This system is also used in hybridization with other cognitive architectures.

The pace of information processing in ACT-R is designed to emulate the human brain and is therefore useful in modeling human behavior, but simulation times are typically accelerated. In its latest form, ACT-R can predict the activation of human brain regions linked to certain behavior and makes use of procedural (how to accomplish a task) and declarative (factual) knowledge (Anderson, 2007; Ritter et al., 2018).

Figure 8.2 shows the architecture of ACT-R. Its components are made up of modules and buffers. Buffers are memories that store content and are

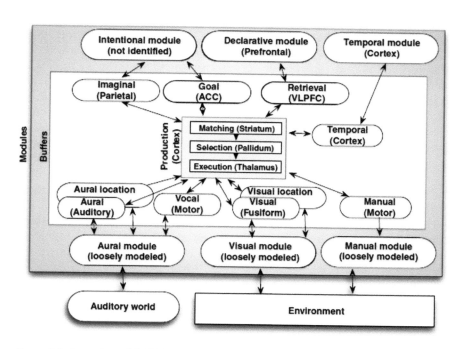

Figure 8.2 Overview of ACT-R architecture. In brackets are the associated human brain components.

Source: Ritter et al. (2018). With permission.

globally visible to all modules. Modules modify and implement information stored in a buffer. There are seven modules in ACT-R, but the four main ones are as follows:

- *Visual module*: identifies objects in the field of view.
- *Goal module*: keeps track of current goal and intents.
- *Declarative module*: retrieves information from buffers (memory).
- *Manual module*: specific for controlling hands.

These modules are mapped to the human brain regions. Modules are active in parallel, but only a single production rule can be executed at a time, taking about 50 ms to fire. The central production module coordinate communication between modules by applying production rules and has access to buffers, updating them through production rules. The visual, auditory, and manual systems are how ACT-R interacts with its environment, such as typing on a keyboard and scanning a computer screen (Ritter et al., 2018).

Learning in ACT-R occurs in several locations and in several ways. A key learning method is declarative memory strengthening: the more a memory is retrieved, the more its activation strength increases.

ACT-R production rules are based on detailed experimental data of memory, learning, and problem solving. ACT-R operates under one goal at a time and its state changes occur through the firing of production rules, which are executed one at a time, but rules can be nested and chained, for example, the output of one rule forming the input to the next rule. The central box in Figure 8.2 is where the system decides which production rule to select next for execution. The matching module identifies the production rule currently possessing the highest "utility", a measure of its relevance to fulfilling the goal, also considering the cost of running that rule: essentially performing a cost-benefit analysis. The cost can be the time it takes to complete the task, and depending on the goal whether it has a time-limited aspect. There is no overseeing system, the system continually updates rules and their cost benefit, and the winner of that update at each rule selection cycle is selected as the next rule to execute.

There is a nuance here that John Anderson believed to be important (Eliasmith, 2013): while the production rules contain symbolic AI concepts, the selection of which rule to execute is not based on symbolic content but instead on a number representing its utility. Anderson placed importance on this subsymbolic aspect of the architecture (a concept shared with neural networks). There are other subsymbolic features of the architecture, a key one is the learning rule based on frequency of firing strengthening future selection of the rule.

Some differences between Soar and ACT-R reported by Ritter et al. (2018) suggest Soar is better for large system AI applications, whereas ACT-R is better for simulating detailed human behavior. Both ACT-R and Soar continue to evolve.

8.2.3 Semantic pointer architecture

Eliasmith (2013) and the team have built models based on the SPA. An attractive feature of SPA is its biological constraints: the model uses neural networks of the neuromorphic type, i.e., spiking neurons, and the responses of neurons are matched to properties of relevant brain regions, the models make use of neurotransmitters, and more. Eliasmith provides an integrated and unified SPA which can solve a variety of problems, without re-training, and this is an advance over many DLNNs that "forget" learned knowledge when re-trained on new problems.

Underlying the SPA is the semantic pointer hypothesis, I quote (Eliasmith, 2013):

- *Semantic pointer hypothesis*: Higher-level cognitive functions in biological systems are made possible by semantic pointers. Semantic pointers are neural representations that carry partial semantic content and are composable into the representational structures necessary to support complex cognition.

In SPA, the semantic pointers are represented by vectors in a high-dimensional space, and this space can represent semantic relationships between representations in an artificial brain (see also Section 8.8). Concepts that are semantically similar lie closer to each other in this space than dissimilar concepts. A high-dimensional vector space is also quite robust to noise. However, Eliasmith says that a method for mapping into this space complex representational structures has not yet been achieved.

Eliasmith makes a distinction between conceptual representations that represent long-term representations of concepts such as behavior of people in contrast with "occurrent" representations which are the active, present activities of the brain. Furthermore, occurrent representations will not carry the full semantic content that conceptual representations possess; occurrent representations are more fleeting. SPA accounts for this through the notion of semantic pointers (a pointer used in the computer science sense), with the advantage that pointers can be manipulated more easily than if operations were performed directly on the content they point to.

For example, in computer programming rather than passing a large data structure to be processed by a function, the data structure remains unmoved at its source and instead a pointer to the data structure is simply passed to the function, so it knows where to find the data. Working with pointers is faster and more efficient. The semantic pointer hypothesis has neural representations that act as pointers: the brain manipulates compact "occurrent" representations which act as addresses to the actual content. Unlike computer science, where pointers can be assigned arbitrarily to content, in the brain model, the relationship between pointer and content has a meaningful connection; they both share in the underlying content semantics, hence the name of the hypothesis. In Eliasmith's hypothesis, the pointer is a compressed version of the content.

The SPA is the model that Eliasmith and his team have developed to encapsulate the ideas around the semantic pointer hypothesis, and bridges the gap between the neural engineering framework, also developed at Eliasmith's lab, that implements biologically realistic neural structures (such as spiking neurons) and the models of cognition being explored.

Eliasmith (2013) associates the basal ganglia in the brain with decision control – this region contains highly interconnected neurons and lies underneath the neocortex near the thalamus. The characteristics of the basal ganglia have been modeled in SPA, including the influence of neurotransmitters on the timing of action selection. SPA, like other cognitive architectures, is a work in progress.

See also Section 8.8, which discusses models related to SPA.

8.3 ADAPTIVE RESONANCE THEORY MODEL OF THE BRAIN

Stephen Grossberg has a long and distinguished research career that crosses three fields: cognitive and systems neuroscience; AI, biological and artificial neural network modeling; and theoretical and mathematical psychology. He is also a prolific writer, and his most recent book *Conscious Mind Resonant Brain: How Each Brain Makes a Mind* (Grossberg, 2021) runs to over 700 pages, and pulls together highlights of his research work. In this section, Grossberg's neural model is discussed: Adaptive Resonance Theory (Grossberg, 2013).

Grossberg divides ART into two streams of activity, each one feeding into the other. The first stream develops ART into a cognitive and neural theory of how our brains attend, learn, recognize, and predict objects and events in a changing world. The second stream takes an engineering mindset to apply ART in real-world applications in engineering, technology, and AI. Starting with the first stream, the foundational hypotheses of ART are supported by psychological and neurobiological experiments, and all the theoretical explanations and predictions it has made have been supported by psychological and neurobiological experimental evidence (Grossberg, 2021).

A key paper "How does a brain build a cognitive code?" (Grossberg, 1980) includes a thought experiment from which ART is derived. The thought experiment considers how the brain can perform error correction without supervision, i.e., autonomously. One of the challenges with many neural network models is they forget trained learning if presented with new training material. This is known as the stability-plasticity dilemma: it is desirable for the model to be plastic, i.e., acquire new information, at the same time to stably remember learned information. This has been somewhat resolved with current large language models that can be fine-tuned and augmented with new information without forgetting (covered in Section 9.4.3). However, catastrophic forgetting is a common problem across many neural network designs, making them only suitable for narrow applications. Carpenter and Grossberg's (1987) ART was designed to learn quickly and stably without catastrophic forgetting.

Grossberg is primarily interested in biological realism and the models are tools with which to test ideas about how our brains make our minds in both healthy individuals and clinical patients. ART and its variants have clarified how, in Grossberg's words "normal brain dynamics can break down in specific and testable ways to cause behavioral symptoms of multiple mental disorders, including Alzheimer's disease, autism, amnesia, schizophrenia, PTSD, ADHD, auditory and visual agnosia and neglect, and disorders of slow-wave sleep". Furthermore, Grossberg says that in the Hebbian rule "cells that fire together, wire together" the computational unit should apply to distributed *patterns* across the network, as implemented in ART, and not individual neuron connections.

For models that aim to have biological plausibility Grossberg (2020) points out the contrast between ART and deep learning, and what is biologically inaccurate with backpropagation in feedforward neural networks compared with ART. Notably, Grossberg says deep learning is untrustworthy because it is not explainable, and unreliable because it can experience catastrophic forgetting. Furthermore, Grossberg (2020) lists 17 biological plausibility problems of deep learning that ART does not have. Perhaps the most pertinent of these is that deep learning uses supervised training while ART uses unsupervised learning. ART models tune their learning to match the statistical variability of the data, and its outputs are explainable. Compared with the lengthy training required in deep learning, ART learns quickly, uses incremental learning without catastrophic forgetting, and can be applied in a rapidly changing environment.

Turning now to the second stream of ART, modeling for engineering applications, Carpenter and Grossberg initially created two ART models: ART 1 was designed for binary input patterns and ART 2 for both binary and continuous, real number input patterns. Input data patterns are represented as vectors. The basic ART unit comprises an input layer (layer 1) holding the activity pattern across a network of feature detectors that respond to the input pattern, and an output layer 2, with adaptive weights attached to connections going from layer 1 to layer 2 and a separate set of adaptive weights attached to connections from layer 2 to layer 1. The layer 1 activity pattern compares, or matches, the bottom-up input pattern, also called an adaptive filter, with the top-down output pattern, also called a learned top-down expectation, from layer 2. A competitive network in layer 2 chooses the cell population that gets the largest input from layer 1 and suppresses the activity of all other cell populations. This is called the winner-takes-all competition. The adaptive weights that abut the chosen layer 2 population learn a time-average of all the bottom-up inputs that are received while the population is active on all the learning trials. As a result, a chosen cell population learns to behave like a recognition category that fires selectively to input patterns that are similar to the ones that trained its adaptive weights.

The ART model clusters input patterns using this unsupervised algorithm. The learned generality of a recognition category is determined by a parameter that is called vigilance. High vigilance leads to learning of concrete or specific categories. Low vigilance leads to learning of general or abstract categories. If a novel input pattern does not fit within an existing cluster, then a new category is chosen and begins to learn that input pattern. Contextually, when new "experiences" (input patterns) are sufficiently unlike existing categories, the model co-opts a free output neuron to learn a new category to represent the pattern, and learning of input patterns stops when all the output (category) neurons are used up, but as ART works with an arbitrarily large number of category neurons, this possibility is easily avoided.

Grossberg notes:

> The vigilance parameter solves a Minimax learning problem; it conjointly uses the minimum amount of memory to achieve maximum generalization. The vigilance mechanism allows ART to learn with the maximum generalization that causes the minimum number of predictive errors. ART does this by a process of match tracking, whereby a predictive mismatch increases vigilance by the minimum amount needed to correct a predictive error, and thus to give up the minimum amount of category generalization with which to do it.

Match tracking and predictive errors occur in ARTMAP systems, wherein two ART systems, one that categorizes input patterns (like multiple visual fonts of the letter A) and the other that processes output predictions (like the predicted answer "A") interact via an associative learning network called the Match Field.

On the question of what evidence there is for vigilance control in the brain, Grossberg and Versace (2008) show that category learning processes mediated by the neurotransmitter acetylcholine can be associated with this mechanism. Grossberg says: "when a big enough mismatch in the nonspecific thalamus activates the nucleus basalis of Meynert which, in turn, broadcasts signals throughout layer 5 of the neocortex, where it releases acetylcholine, that causes vigilance to increase".

When category nodes (existing clusters) and feature nodes (input patterns) interact, they either resonate (i.e., match within the vigilance radius) or there is a mismatch, which results in a reset and search, whereby the next input pattern is processed. The "adaptive" label in ART refers to the fact that, when fast bottom-up input patterns and top-down expectations match well enough to generate a sustained resonance between layers 1 and 2, then fast learning is triggered in both the bottom-up and top-down adaptive weights. Unlike the perceptron, which learns from the difference, or *mismatch*, between an output and a desired output (the target), in ART category, learning is triggered when a good enough *match* occurs. If a match is not found between a novel input and an output, the algorithm searches for another cell population that starts to learn a new category about that input. As input trials continue, ART bottom-up adaptive filters and top-down expectations learn to select and pay attention to the *critical features* that control predictive success, while inhibiting irrelevant outlier features, thereby also reducing noise. This two-level ART module is a basic unit and units can be stacked to perform more complex learning tasks.

The mathematical analysis of ART was carried out in a series of papers, e.g., Carpenter and Grossberg (1987) proved a complete set of mathematical theorems that characterizes how the ART 1 model processes arbitrarily binary input patterns, each with arbitrarily many features, into winner-take-all, stable learned categories. Variants and generalizations of ART 1 include ART 2, which can learn to classify continuous or binary input patterns, ART 2A, ART 3, fuzzy ART, fuzzy ARTMAP, distributed ARTMAP, Gaussian ARTMAP, predictive ART, and more. Carpenter and Grossberg showed that an ART algorithm can learn an entire database in one to five fast learning trials, and stably recall the learned categories. While ART is unsupervised, the various ARTMAP variants can be learned using arbitrary combinations of unsupervised and supervised learning.

The original computer simulation work with ART in the 1980s–1990s was developed using continuous value, or rate-based neuron outputs, rather than using the more realistic spiking neuron outputs, due to limitations of computers of the day. However, as computer speeds and capacities increased, it was shown how all the many models developed by Grossberg et al., including models of perception, learning, cognition, emotion, and action, could be recast with spiking neurons. This was possible because all these models used realistic neurophysiological membrane equations to describe neuron activations. For some examples of such spiking models, see Grossberg and Versace (2008), Cao and Grossberg (2012), and Pilly and Grossberg (2013). One of the most recent ART models, predictive ART, or pART (Grossberg, 2018), unifies in one model neural architecture many of the individual models that had initially been published to explain and simulate on the computer different psychological and neurobiological databases. The pART architecture includes, for example, the cognitive-emotional-motor (CogEM), model of how cognition and emotion interact. CogEM can accordingly explain experimental data on how thoughts and feelings interact (Grossberg, 2018). All Grossberg's archival articles can be downloaded from his web page: sites.bu.edu/steveg/.

The transfer of ART into large-scale applications in engineering, technology, and AI has been ongoing throughout its development. Some of these have been written up, e.g., application of ART 1 for the identification and retrieval of engineering designs at Boeing (Smith et al., 1997). For a survey of ART models in engineering applications, see Brito da Silva et al. (2019). Many of these applications are a foundation part of commercial intellectual property and remain confidential. ART has also been implemented in hardware, as illustrated by a book on developing ART VLSI chips by Serrano-Gottarredona et al. (1998).

To fully understand how ART works the reader is encouraged to examine the algorithm equations and to run code. See, for example, the software repository of multiple models created by Gail Carpenter and her students at http://techlab.bu.edu/resources/software/C51.html, as well as Hagan et al. (2014) who also has a website with MatLab code for ART 1, and other ART source code examples on GitHub.

8.4 HARMONIC OSCILLATOR RECURRENT NEURAL NETWORKS

The occurrence of rhythmic electrical patterns in the brain has been examined for some decades (Buzsáki, 2011), but such oscillatory dynamics and the consequence of axon conductance delays have been dismissed as epiphenomena by some neuroscientists. New research by Wolf Singer and his research team (Effenberger et al., 2023) has revealed an intriguing role for these phenomena in computation.

To test whether oscillatory dynamics influence performance of the brain's computation, the phenomenon was simulated in a recurrent neural network (RNN) model and applied to a standard pattern recognition benchmark, MNIST (Le Cun et al., 2010). Unlike the typical integrate-and-fire model of neurons in artificial neural networks, Effenberger et al. (2023) configured a network node on a mesoscale, to be thought of as representing a population of neurons and characterized by damped harmonic oscillation (DHO) – their model was thus named harmonic oscillator RNN (HORN). The benchmark simulations showed that HORN significantly outperformed RNN in parameter efficiency, task performance, learning speed, and noise tolerance. Simulations were run to test cortical-related features in HORN including nodes with different oscillation frequencies, different conduction velocities for the recurrent connections, synapses with unsupervised Hebbian learning, and recurrently connecting two HORN networks in a multilayer architecture. All these biologically inspired additions further improved the benchmark test performance.

The HORN nodes have a local memory which is populated by Hebbian learning and supports parallel search of input patterns across the memory.

Just a single DHO node has computational capabilities, before considering a network of such nodes, for example, it can encode stimulus information in the oscillation phase, and self-sustained oscillations are possible through the recurrent feedback connections. Nodes can resonate at fractional harmonics of their natural frequency. The natural frequency parameter of a DHO acts as a band filter that selects and amplifies input signals, modulating their amplitude, frequency, and phase in a non-linear way. Each DHO node turns an input pattern into a harmonic oscillation, leading to a rich set of learning behaviors, described by Effenberger et al. (2023) as: "fading memory, resonance, entrainment, fine-scale synchronization, phase shifts, and desynchronization both on the single node and on the network level which can be exploited for learning".

Effenberger et al. (2023) first modeled a homogenous HORN network with no conductance delays and identical oscillation parameters for all nodes. The network was fed the 2D images in MNIST as a time series by reading intensity values in scan-line order from top left to bottom right of the images. The network was trained using a supervised backpropagation-through-time algorithm, was balanced with excitatory and inhibitory outputs, and was initiated with random weights. The homogenous HORN model was compared with traditional RNN architectures, and the test performance results are shown in Figure 8.3 – homogeneous HORN^h is in blue. The HORN model outperformed the other RNN models in speed to train the model with high accuracy, achieved the highest test accuracy, and had superior performance robustness with input noise.

The model parameters were next varied to reflect physical cortical network heterogeneity, so each node had a different natural frequency, damping coefficient, and excitability parameter from its neighbors. The performance test

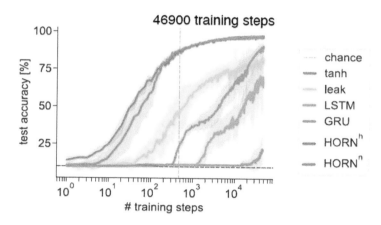

Figure 8.3 Test accuracy (y-axis) of various RNN models trained on the MNIST as time series data, as a function of the training steps (x-axis).

Source: Effenberger et al. (2023). With permission.

result for the heterogeneous HORN^n is shown in Figure 8.3 in orange and shows performance comparable with the homogeneous model. However, further simulations between the two revealed the heterogeneous model to have superior robustness to noise and was better at simultaneously holding information about temporally segregated stimuli.

Cortical neuron axon conductance is of the order of 0.5 to 10 m/s which results in distance related coupling delays between neurons. Effenberger et al. (2023) reflected these delays by uniformly distributing coupling delays across all the recurrent connections in homogeneous HORN^h and in heterogeneous HORN^n models from within a distribution $[1, d_{max}]$. Increasing maximum delay factor d_{max} which increases delay heterogeneity also has the effect of increasing the task performance accuracy. The heterogeneous model also outperformed the homogeneous model.

The multilayer combination of HORN networks was tested with two homogeneous HORN^h models (layer 1 and layer 2) connected with feedforward and feedback connections. The input patterns were presented to layer 1 and the output was read from layer 2. Each model consisted of 32 nodes and was implemented with sparse reciprocal connections between the layers. The number of trainable parameters varied, and the test accuracy is shown in Figure 8.4. The two-layer model outperforms the layer 1 model, achieving high accuracy with fewer training steps.

Based on these results Effenberger et al. (2023) postulate that networks of coupled oscillators performing massively parallel analog computations can relate many input variables and solve complex cognitive tasks and do so at speed despite the relative slowness of neurons.

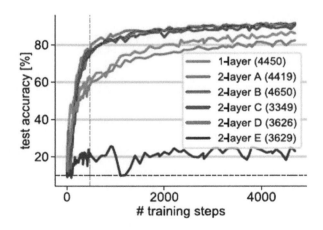

Figure 8.4 Test accuracy of one layer and two-layer HORN^h homogeneous models with different number of trainable parameters (in parenthesis).

Source: Effenberger et al. (2023). With permission.

8.5 NUMENTA AI MODELS

8.5.1 Introduction

Numenta is a research company led by neuroscientist Jeff Hawkins with the mission to apply neuroscience discoveries to AI. Numenta has developed HTM (Hawkins and Blakeslee, 2014; George, 2008; George and Hawkins, 2009), a model for how the brain might work. Numenta's research team led by Hawkins continues to develop HTM and the broader theory that supports it, called thousand brains theory of intelligence (Hawkins, 2021; Hole and Ahmad, 2021).

8.5.2 HTM and thousand brain theory

HTM is a specific realization of thousand brains theory (Hawkins, 2021). A detailed description of the first version of HTM is given in George's PhD thesis (George, 2008), what follows is a summary of more recent versions (Hole and Ahmad, 2021). Numenta also keeps a continually updated, "living book", on HTM which it calls biological and machine intelligence (see Hawkins et al., 2020).

HTM is a simplified model of the neocortex, comprising a uniform layout of HTM neurons divided into vertical (major) columns each of a uniform structure of six horizontal layers, the bottom five contain mini-columns of HTM neurons – these mini-columns can cross the horizontal layers. Neurons in a mini column connect to other neurons. The theory applies the same common "cortical" algorithm across all the neurons. HTM regions (comprising the major columns) connect in approximate hierarchies, an example is depicted in Figure 8.5.

An important part of the HTM model is storage of data by neurons using a sparse distributed representation (SDR), reflecting the sparse activity patterns in the human brain: the percentage of active neurons at any given time is around one to a few percent.

The ideas in SDR can be traced back to Pentti Kanerva's sparse distributed memory (Kanerva, 1990). SDR uses binary vectors to indicate active and inactive neurons, where 1 indicates active. First-generation implementations of HTM had neurons carry continuous signals with discrete time steps; however, third-generation models introduced the concept of dynamic time, allowing neurons to communicate in spikes.

SDR has powerful properties, for example, SDR is robust to noise (i.e., a few inaccurately firing neurons have little impact on the result) and has vast capacity. For more on SDR, see Ahmad and Hawkins (2015); Ahmad and Hawkins (2016); Purdy (2016); and Cui, Ahmad, and Hawkins (2017).

The capacity of SDR is easily demonstrated (Ahmad and Hawkins, 2015). The activity of neurons in the HTM model is given by their axon state: 0 or 1. The activity state of the HTM model can be represented at any given

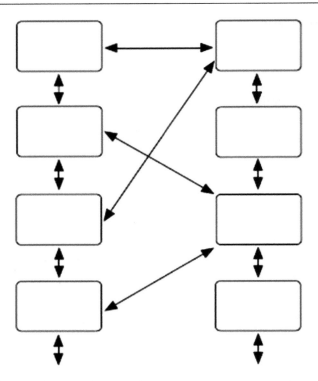

Figure 8.5 HTM columns: two approximate hierarchies of vertically connected regions with horizontal connections between the hierarchies.

Source: Hole and Ahmad (2021). With permission.

time by collecting all the neuron output states into a vector, for example, this small HTM model has 40 neurons in some point in time we call X:

Output_state_X = [0000100000100000001000000010000000000010]

This model is sparse with only five neurons fired up at a given time. Larger HTM models set three configuration parameters as follows:

- n: the length of the SDR vector, which can be 1,024 to 65,536.
- w: the number of bits in the vector that can be on, which can be 10 to 40 (reflecting sparse activation).
- s: the sparsity of the model, given by s = w/n, and can be 0.05%–2%.

A typical HTM model has n = 2048 and w = 40. While the number of possible HTM states is a lot less than the total possible states of a binary vector of length n, which is 2^n, this does not matter: the actual number of states is given by the combination formula from elementary probability theory, i.e., how many ways can you select w objects from a collection of n objects (and the order of picking up w is not important). So, for n=2,048 and w=40 the combination formula gives the number of possible states as approximately 2.37×10^{84}, greater than the number of atoms in the observable universe. In the human brain n=80–100 billion, so on this metric, the capacity of the brain is limitless.

The equivalent of sensory organs in HTM are called encoders, these convert various data types: dates, times, and numbers, such as coordinates, into SDRs. Figure 8.6 shows how the HTM works at a high level.

In Figure 8.6, definition of terms are as follows

- *Proximal* is a feedforward input from a lower layer neuron (a dendrite forward connection).
- *Apical* is a feedback input from a dendrite backpropagation and can be from any layer, given the possibility of cross-connections between nearby columns.
- *Basal* is an input from neurons within the same layer (a dendrite "forward" connection from within a layer).

The three types of input signals are first summed (integrated) before applied to the HTM neuron. The HTM neuron models the electric potential state of real neurons by having three states: active, predictive, and inactive. Only proximal input signals result in switching the HTM neuron into an active state whereby it outputs a "1" signal in its axon output, but only if it is currently in the predictive state. In this model, the output is a continuous signal in time and not a spike. Signals in the apical and basal inputs put the HTM neuron into a predictive state.

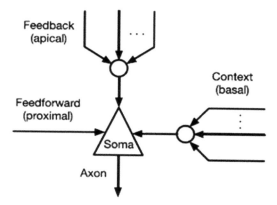

Figure 8.6 HTM neuron sketch: feedforward input determines whether the soma moves into active state and fires a signal on the axon. The sets of feedback and context dendrites determine whether the soma moves into the predictive state.

Source: Hole and Ahmad (2021). With permission.

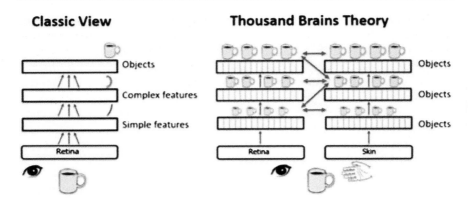

Figure 8.7 Rather than learning one model of the world, the Thousand Brains Theory of intelligence states that every part of the neocortex learns complete models of objects and concepts. Long range connections in the neocortex allow models to work together to create the perception of the world.

Source: https://www.numenta.com/blog/2019/01/16/the-thousand-brains-theory-of-intelligence/With permission.

The term "Thousand Brain" in Numenta's theory refers to each cortical column (major column), learning complete models of objects, Numenta's view is that the brain creates thousands of models simultaneously distributed across the neocortex, so that many columns are modeling *the same object*. This is in contrast with deep learning, where an object is recognized just once at the output layer, see Figure 8.7 comparing deep learning and thousand brain theory.

Numenta proposes concepts such as the memory mnemonic technique called method of loci, also known as memory palace, where an object that we wish to remember is linked with or "placed" in a familiar location, so with the brain Numenta proposes memories are associated with spatial locations through grid cells, a type of neuron they suggest is present throughout the neocortex (Hawkins et al., 2019).

HTM is suitable for sequence learning, and as the model in Figure 8.6 shows, it shares the recurrence feature of recurrent neural networks, and has led Numenta to market HTM for such applications. Training of HTM neurons is performed by a local, unsupervised Hebbian learning rule, i.e., neurons that fire have their connection weights strengthened, inactive neurons have their weights weakened. In HTM weights are also binary, 0 or 1, which simplifies the model. Learning is performed in specific dendrites and new synapses are added, or existing ones purged depending on their activity. HTM neurons predict their output and if accurate their connections are strengthened. A never before encountered pattern will not be predicted and the algorithm will select the neuron with the closest match in each mini column to learn the pattern (Hawkins et al., 2017).

HTM has a continuous learning capability that performs in real time, there is no offline training – the training is continual online. Another major difference with deep learning is the reduced number of hyperparameters, which require fine tuning, largely a result of HTM's simple learning rule in contrast with machine learning (ML) backpropagation.

Numenta has performed successful experiments with HTM in computer simulations (Cui et al., 2016). Billaudelle and Ahmad (2016) ported HTM onto a neuromorphic computing platform called hybrid multi-scale facility (HMF), which has at its core a neuromorphic processor, called high input count analog neural network (HICANN). HMF uses mixed signal analog and digital modeling with spiking neurons. The research had mixed results; the greatest challenge was adapting HTM learning rules to the native plasticity features available on HMF. HICANN is an application-specific integrated circuit (ASIC) with its rules hard-wired, creating a mismatch if the fixed rules cannot represent exactly the desired algorithm.

A more successful implementation was reported by Zyarah and Kudithipudi (2019) who were able to accelerate HTM in a neuromorphic architecture implemented in AMD Xilinx field programmable gate array (FPGA) processors. FPGAs are highly flexible, implementing in hardware directly algorithm logic, thereby achieving the speed of hardware with a precise representation of the software. The authors achieved a 1,364 factor speedup over a purely software simulation and suggested this architecture as a basis for a full hardware implementation of a machine governed by HTM. In further research by Zyarah, Gomez, and Kudithipudi (2020), they propose a memristor-based mixed signal HTM architecture that includes spatial and temporal aspects of HTM.

In more recent research at Numenta, Hunter, Spracklen, and Ahmad (2022) modeled two aspects of sparsity that exist in the human brain: sparse interconnections between neurons, and sparse activity of neurons. Using this model, they demonstrate a factor of 100 improvement in throughput and energy efficiency running on Xilinx FPGAs.

See the extensive Numenta literature for more in-depth theory, analysis, and results of HTM simulations on its web site.

8.6 DEEP LEARNING NEURAL NETWORKS

8.6.1 The origins of generative AI

DLNNs were mentioned at the end of Chapter 2. We pick up the story here for what is relevant in our mission. As we described earlier, the rise of deep learning was sparked by the huge reduction in training times that state-of-the-art GPGPUs made possible. Large and deep-layered models with millions of neurons could be trained in hours or days instead of many weeks and months. This accelerated research work and when the deep learning teams entered their algorithms in vision competitions, they beat all incumbent algorithms that were based on decades of expertise in domain-specific vision science. DLNNs, the general-purpose algorithm honed on data and requiring no expertise in the application domain was the way forward.

Today, these GPGPUs are just called GPUs with specific product lines aimed at AI algorithm acceleration. At the time of writing top end GPUs dominate the market, Nvidia being a key player, for helping train DLNNs, typically in data centers, with competition from specialist processor makers such as Cerebras and Graphcore. There are examples where the latest generation CPUs are sufficient for training models. For inference mode, the market is more fragmented, GPUs can be used, field programmable gate arrays (FPGAs) have their advantages, as do many specialized ASIC chips, and the CPU makers continue to enhance their designs to support inferencing on a CPU.

Deep learning supported by accelerator chips gave rise to ML algorithms and applications crossing over from research laboratories to the enterprise and vertical industries such as finance, healthcare, telecoms, engineering manufacturing and more. However, while ML applications can be smart in highly focused applications, they are not intelligent at human levels. Meanwhile, the deep learning researchers in universities and the high-tech research arms of the major tech giants, notably Google, Microsoft, and Meta, continued intense research activity, and there was a freedom of exchanging ideas to a degree I have rarely witnessed: research paper pre-prints appeared rapidly on arXiv.org and often with shared code on GitHub.

A major step change in deep learning design took place with the publication of Ian Goodfellow et al.'s generative adversarial nets (GANs) in 2014, by Yoshua Bengio's team at Universite de Montreal. An important feature of the GAN was the interplay between two neural networks: the generator (a pupil) and the discriminator (a teacher). The generator attempts to fool the discriminator by generating plausible output (e.g., maybe it is being trained to create a Picasso-like image, and the training data is all Picasso's artwork). The discriminator, who has been trained to recognize the genuine item, assesses the output of the generator and penalizes it if recognized as fake. This training process ends when the discriminator cannot distinguish a real Picasso from the generator output. At this point the discriminator is discarded, and inferencing continues with the generator.

GANs led to more research interest and funding, and an explosion of applications and public interest in generative AI. In December 2015, Sam Altman, Greg Brockman, Reid Hoffman, Jessica Livingston, Peter Thiel, Elon Musk, Amazon Web Services (AWS), Infosys, and YC Research announced the formation of OpenAI with $1 billion of support.

The next key moment in this very brief history was the publication of a seminal paper announcing a new deep-learning architecture for natural language processing: the transformer. With the confident title

"Attention Is All You Need" by Vaswani et al. (2017) from the Google Brain research group, the paper introduced an architecture shift in design by dropping the use of an RNN in favor of relying entirely on an attention mechanism. A significant benefit of this architecture is that it can be massively parallelized, which is perfect for training on GPUs, unlike RNNs. The attention mechanism is a way of giving greater weighting to key words that appear before each word in a text sequence, and depending on the application and model architecture, can also make use of key words that appear after each word in a text sequence. Vaswani's paper applied the transformer to language translation and used an encoder and decoder structure.

This approach was picked up by the OpenAI researchers and it published the first generative pre-trained transformer (GPT) paper the following year: "Improving Language Understanding by Generative Pre-Training", Radford et al. (2018). GPT has 110 million parameters, where parameters are mostly the weights connecting neurons and which are tuned during training, so number of parameters gives an indication of the number of neurons in the model. The following year 2019, OpenAI released GPT-2 with around 1.5 billion parameters, and in 2020 released GPT-3 with 175 billion parameters. So far, the interest in these models, and other large language models (LLMs) being developed by teams at Google, Meta, and others, was confined within the AI research community. On November 30, 2022, OpenAI launched to the public ChatGPT in what it assumed was just another trial to gauge interest. Within a couple of months the public appetite for ChatGPT grew astonishingly, reaching 100 million users per day, and it is fair to say that a new era in AI was launched.

The first ChatGBT was based on GPT-3.5, an LLM trained with data available on the internet at the training time of September 2021. The date to which GPT-4 has been trained has not been divulged by OpenAI. There is a marked difference in how GPT3/3.5 and GPT4 are trained: in the former, one agent scours the internet for training data, whereas in the latter, multiple agents explore different parts of the internet and then share their findings, a method known as mixture of experts.

Using techniques such as fine-tuning and retrieval augmented generation, it is possible to bring a large foundation LLM (also called frontier model) up to date with recent information. The parameter size of GPT-3.5 and GPT-4 has also not been revealed by OpenAI (it is believed to have 1.7 trillion parameters).

The capabilities and performance of GPT-4 are best seen with reference to Figure 8.8 from the OpenAI paper on GPT-4 (OpenAI, 2023).

Figure 8.8 shows the performance of GPT-3.5 in various academic and professional exams in blue and the improvements using GPT-4 in green. The most remarkable is the uniform bar exam performance that shot up from around 10% GPT-3.5 to around 90% GTP-4. OpenAI says there was no specific training performed for taking the exams, as shown in Figure 8.8.

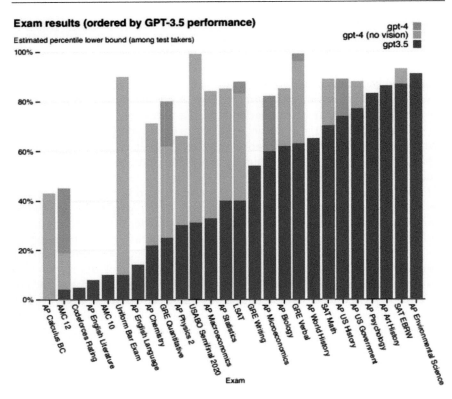

Figure 8.8 The performance of GPT-4 quoted in the GPT-4 technical paper by OpenAI.
Source: OpenAI (2023). With permission.

The GPT technology and LLMs built with it will have a profound impact on the spread of AI into society, and the next five years will see major changes as businesses across many verticals take measure of how it will affect their processes and mode of business, and there will be changes on the job front, as some will disappear, and new ones appear requiring knowledge of working with AI systems.

8.6.2 LLM emergent properties

At the outset, it is worth clarifying that LLMs are black boxes under the hood. The AI researchers who build LLMs are experts in the model architecture and train these models to achieve the highest performance, but the result is still a network of artificial neurons connected in various configurations and attenuated by connection strengths. How the LLM model produces its excellent results is unknown. In an effort to better understand how an LLM produces its results, AI researchers have treated the models as experimental subjects and conducted various tests on them, reported here.

Generative AI/LLM models are excellent for answering questions and writing human-like quality essays, with knowledge imbibed in the machine through training. In their natural language understanding, they offer vastly superior man-machine interface capabilities than previously available. Moreover, these LLMs also show emergent properties that deserve closer understanding and could place this research on the path to an engineered HLAI, albeit at a very early stage. Researchers mainly from Google and Stanford published a paper in October 2022, "Emergent Abilities of Large Language Models" that shows intriguing properties of LLMs (Wei et al., 2022).

The first question to ask is what an emergent property is exactly. The authors of this paper suggest this definition: "Emergence is when quantitative changes in a system result in qualitative changes in behavior". Through scaling up, seeing an unexpected result revealing a new capability would be classed as an emergent property. Going back to the paper, the specific emergent property the authors consider is one that is not present in smaller models but is present in larger models.

To set the stage consider the Beyond the Imitation Game benchmark (known as BIG-Bench), consisting of 204 tasks contributed to by 132

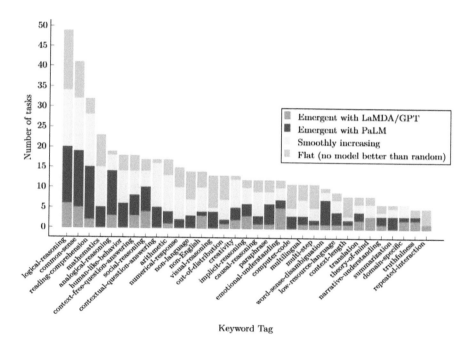

Figure 8.9 Emergent abilities, for example in PaLM performing logical-reasoning tasks. The y-axis shows tasks correctly completed across multiple benchmarks (including BIG-bench). "Smoothly" and "Flat" show no emergent ability, that is, model scaling predicts the results.

Source: Wei et al. (2022). Copyright acknowledged.

institutions (Srivastava et al., 2022). The BIG-Bench evaluation suite has been designed to be beyond the capabilities of current language models. Figure 8.9 shows on the x-axis the bucketing of the 204 challenges: on the left most there are just under 50 logical-reasoning problems (the most numerous category) and the right most is "repeated interaction" with five examples. The y-axis provides two pieces of information: the number of tasks in the bucket, and the performance of those tasks shown in color code, where pink represents the tasks that were no better than random results, beige represents good performance that scaled with the size of the model, and the shades of blue represent emergent properties that the authors discovered. For the blue cases simply extrapolating from properties of smaller models would not have revealed the emergent result observed in larger models. Results are shown for several LLM models: OpenAI GPT, Google PaLM and an earlier Google model, LaMDA.

To better assess the emergent properties, see Figure 8.10. Experiments were performed running the BIG-Bench suite across LLM models with various parameter sizes. We see in the results that performance is random until a critical parameter size point is reached and then performance in various BIG-Bench categories suddenly starts to improve. For modular arithmetic GPT-3, for example, accuracy starts to improve from 10 billion parameters upwards. While the 30% accuracy at 100 billion parameters is not

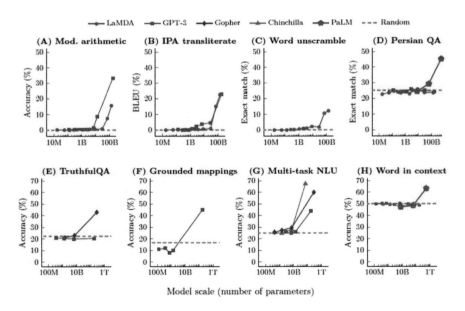

Figure 8.10 Emergent properties appear in LLMs as the parameter size (x-axis) scales up.

Source: Wei et al. (2022). Copyright acknowledged.

impressive by human standards, we need to understand these are benchmarks designed to try to fail language models – they are hard challenges, so even 30% for a machine not designed for mod arithmetic is good. These results are emergent because they are unexpected and reveal a hidden capability that emerges at high scale. In a follow-up blog, Jason Wei has catalogued 137 emergent abilities of LLMs (Wei, 2022).

LLMs are built with self-supervised algorithms. Where supervised algorithms are provided with targets (the known solutions), and unsupervised algorithms have no targets of any form, self-supervised algorithms use the input data to help drive the algorithm, so self-supervised is more like unsupervised than supervised algorithms. A significant advantage of not using supervised algorithms is that it removes the task of creating labeled targets, instead the LLM can be trained on raw data from multiple sources with little preparation.

LLMs have pre-trained models and in query mode use fine-tuning and augmentation to provide the best responses, including in specialized domains with augmentation of data from, for example, vector databases. LLMs used in this way are known as foundation (Bommasani et al., 2022) or frontier models, and this approach is called transfer learning. Mining the subsequent system is called prompting or prompt engineering, including iterative prompting, which has grown as the art of querying LLMs. It is also known as chain-of-thought reasoning with the LLM (Suzgun et al., 2022; Wei et al., 2023), where a few chain of thought demonstrations are provided as exemplars in prompting, improving the ability of LLMs to perform complex reasoning. Jason Wei and the team (2023) have performed experiments on three LLMs and show that chain-of-thought prompting improves performance on a range of arithmetic, commonsense, and symbolic reasoning tasks. Prompting a PaLM model (540 billion parameters) with eight chain-of-thought exemplars achieved high accuracy on the GSM8K benchmark (Cobbe et al., 2021) of math word problems.

Another dramatic emergent property of GPT was revealed in a paper by Power et al. (2022) who found that overfitting on an algorithmic data set yielded a generalization capability. The OpenAI team trained a small GPT model with only 400k parameters and trained to perform the binary operation of division mod 97 with only 50% of the data in the training set. The result is shown in Figure 8.11, where the red curve shows the training performance and the green curve shows performance on a validation data set (i.e., data not used in training, the validation is performed after each training step). While training accuracy is achieved at around 1,000 iterations, the model shows poor learned performance until the model is overfitted to the point of 100,000 iterations when suddenly the validation performance starts to rise and achieves 100% accuracy on the validation data at a million training cycles.

A paper published by Bubeck et al. (2013) from Microsoft Research investigated an early version of GPT-4 and found it capable of logical reasoning.

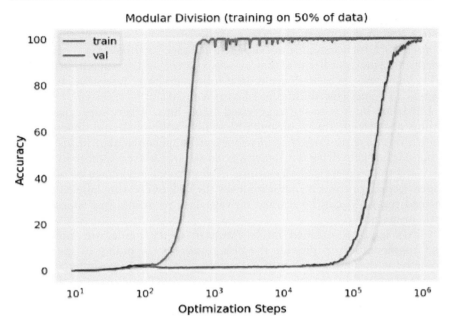

Figure 8.11 Overtraining a small GPT (400k parameters) revealed an emergent general-ization capability.

Source: Power et al. (2022). Copyright acknowledged.

Figure 8.12 shows an example of the experiments conducted, in this case, asking the system to stack a set of objects in a stable tower. A comparison is also made with ChatGPT. The result shows GPT-4 possessing common sense in the suggested solution (an alternative, perhaps human attempt, would be to switch around the book and laptop, with the laptop as base, given the laptop is typically heavier than a book).

A paper by Iddo Drori et al. (2022) from MIT and other institutions report a method for solving questions from an advanced mathematics problem dataset designed to test mathematical reasoning. The dataset comprised the MATH benchmark and questions from MIT university courses. MATH includes questions on Algebra, Counting and Probability, Intermediate Algebra, Number Theory, and Pre-calculus, while the MIT courses included calculus, differential equations, and linear algebra.

The authors tested OpenAI Codex, an LLM model pretrained on text and then fine-tuned on code using code-davinci-002 vector space. Problems were randomly selected from the dataset but excluded any that contained input images or required proofs, and the results showed Codex was able to automatically answer correctly 81% of these university-level mathematics problems with few-shot learning, and 70% accuracy with zero-shot

GPT-4 vs ChatGPT

Prompt: Here we have a book, 9 eggs, a laptop, a bottle, and a nail. Please tell me how to stack them onto each other in a stable manner.

GPT-4:

One possible way to stack the objects onto each other in a stable manner is:

- Place the book flat on a level surface, such as a table or a floor. The book will serve as the base of the stack and provide a large and sturdy support.
- Arrange the 9 eggs in a 3 by 3 square on top of the book, leaving some space between them. The eggs will form a second layer and distribute the weight evenly. Make sure the eggs are not cracked or broken and handle them gently to avoid dropping them.
- Place the laptop on top of the eggs, with the screen facing down and the keyboard facing up. The laptop will fit snugly within the boundaries of the book and the eggs, and its flat and rigid surface will provide a stable platform for the next layer. Make sure the laptop is closed and turned off, and do not press any keys or buttons.
- Place the bottle on top of the laptop, with the cap facing up and the bottom facing down. The bottle will add some height and balance to the stack, and its cylindrical shape will reduce the contact area with the laptop. Make sure the bottle is empty and closed, and do not shake or tilt it.
- Place the nail on top of the bottle cap, with the pointy end facing up and the flat end facing down. The nail will be the final and smallest object in the stack, and its sharp tip will prevent it from rolling or sliding off the bottle. Make sure the nail is clean and straight, and do not touch or move it.

ChatGPT:

It would be difficult to stack all of these objects on top of each other in a stable manner without using any extra objects. The book and laptop are likely too large and heavy to stack on top of the eggs without damaging them. The bottle and nail may be too small to provide a stable base for the other objects.

Here is one possible way to stack the objects:

- Place the bottle on a flat surface.
- Carefully balance the nail on top of the bottle.
- Place the eggs on top of the nail, making sure they are balanced and not tilting to one side.
- Place the laptop on top of the eggs, again making sure it is balanced and not tilting to one side.
- Finally, place the book on top of the laptop.

This stack may not be very stable, so it is important to be careful when handling it. Alternatively, you could try stacking the objects in a different order or using some of the objects as support for others to create a more stable stack.

Figure 8.12 An exhibit from experiments with GPT-4 demonstrating common sense.

Source: Based on Bubeck et al. (2023). With permission.

learning – solutions were generated programs that solve the math problems. This work not only solves high-level mathematics course questions at a human level, but it also explains solutions and generates university-level mathematics course questions at scale in a matter of seconds.

The term "zero-shot learning" is used in the research literature to mean testing problems on a model from examples which were not used to train the model – the model has never seen these problems before. In the Drori paper, zero-shot learning involved providing either just the original question or adding brief additional instructions such as "write a program", "using SymPy", or "using simulations".

To understand few-shot learning, the concept of a vector embedding space needs to be explained. Information to be added to the pre-trained model can be performed without re-training the model, which is an expensive operation. Instead, vector embedding spaces can be added to the model. Vector embedding comprises thousands or millions of examples that are coded into numerical vectors using ML, and then this space is searched by the LLM algorithm when generating a solution. OpenAI has released this option as an embedding endpoint to use with its models (Neelakantan et al., 2022), and the user needs to specify which vector space is to be used. Drori et al. (2022) use GPT-3 with text-davinci-002 vector space and Codex with code-davinci-002 vector space.

Few-shot learning can be performed when zero-shot learning is unsuccessful. The method used by Drori et al. (2022) was as follows: *all* the questions were embedded using OpenAI's text-similarity-babbage-001 embedding engine (a 2,048-dimension vector space). Given a test coding question, the most similar solved questions to this test question are calculated (the authors use the cosine similarity embedding metric) and the code solutions are extracted to be used as examples to accompany the test question.

Again, remember the test question has not been shown with its solution to the model, the only action performed with the test question is its conversion to the vector space to find similar questions which do have code solutions embedded in the vector space. So, from the extracted similar question-code couplets one example is selected to accompany the test question shown to the model, and if the problem is not solved, this is repeated a maximum of five times.

These examples demonstrate capabilities of deep learning models beyond previous expectations and transferring these models to applications in the enterprise and across industries and education will be impactful.

Finally, using LLMs is not without risk, as a recurrent issue is false information being generated, charitably described as "hallucinations" within the AI community (Ganguli et al., 2022) – AI pioneer Geoffrey Hinton prefers the term "confabulation", a technical term in neuropsychiatry for innocently created false memory; Hinton likens this feature of LLMs as similar to human imagination (Hinton, 2024). A worrying emergent feature has been found by Parrish et al. (2022) that increasing parameter size in LLMs creates more social bias in the output.

8.7 BIOLOGICALLY PLAUSIBLE MODELS

8.7.1 Backpropagation in the brain

The backpropagation method for updating the weights in an artificial neural network was originally considered as biologically unrealistic but given its success with deep learning, it is being re-appraised in attempts to model computation in the cortex (Whittington and Bogacz, 2019). The key step is the backpropagation step where downstream neurons send corrective information back upstream, this information being based on a gradient that drives changes in synaptic weights to reduce error in pattern matching. The neuroscience behind top-down feedback is now accepted, see e.g., Gilbert and Li (2013). In a review of attempts to apply backpropagation to the brain by Timothy Lillicrap et al. (2020), the view is that top-down feedback connections may induce neural activities whose differences can be used to locally approximate backpropagation feedback signals and drive learning in deep neural networks in the brain. In an earlier paper by Timothy Lillicrap et al. (2014), the authors showed that a simple deep learning algorithm which evaluates weight corrections by multiplying error signals by random synaptic "weights" still allows the network to learn and extract useful information from feedback signals. Synapses in the brain communicate information unidirectionally, and so feedback of upstream weight information, as required in backpropagation is biologically implausible. Using random weights for the feedback process, the authors demonstrate that the network learns to learn, and with an accuracy equal to traditional backpropagation. This method demonstrates a biologically plausible basis for neurons exploiting error signals generated in upstream locations in the brain.

The paper by Yuhang Song et al. (2020) has proposed an algorithm that overcomes the limitations of backpropagation as a biologically realistic algorithm and instead has created a brain learning (BL) algorithm that addresses three crucial requirements

- BL produces the same updates of the neural weights as backpropagation.
- Weight modifications are local and can be performed in parallel.
- BL can be modified to operate fully autonomously.

Song et al. suggest that this algorithm is suitable for implementing in neuromorphic processors.

An alternative approach with the same aim of applying a form of back-propagation to the brain is given by Joao Sacramento et al. (2017), who propose that the errors at the heart of backpropagation are encoded on the distal dendrites of cross-area projecting pyramidal neurons. In this model the corrective errors result from the mismatch between lateral interneurons and the top-down feedback from downstream cortical areas. Synaptic learning in bottom-up connections is driven by these error signals flowing through the dendrites. To quote the authors:

> Cross-area synapses onto the dendritic compartments learn to reduce the prediction error between the somatic potential and their own dendritic branch potential. Additionally, lateral synaptic input from local interneurons learns to cancel top-down feedback from downstream brain areas. When a new top-down input arrives at distal dendrites that cannot be matched by lateral inhibition it signals a neuron-specific error (encoded on the dendritic potential) that triggers synaptic learning at a given pyramidal cell. As learning progresses, the interneurons gradually learn to cancel once again the new input until eventually learning stops.

In the paper, the authors say this model is identical to learning by backpropagation.

The Equilibrium Propagation (EP) is a learning framework introduced by Benjamin Scellier and Yoshua Bengio (2017) for energy-based models which do not suffer from some of the biological implausibility issues of backpropagation. In EP the same computational circuit is used for feed forward (creating a prediction) and feed backward (training to reduce errors). In this model, errors are reduced by driving the model's energy function to a minimum. The nodes (which can represent neurons or neuron circuit ensembles) gradually move toward lower energy configurations. In the first phase, feed forward prediction, the inputs are clamped, and the network relaxes to a local energy function minimum. In the second phase, target values are compared with the outputs and the outputs are nudged toward their targets, reducing the prediction error.

The EP model is not without its challenges, such as requiring symmetric weights between the network nodes: symmetric connections mean using the same weights for forward and backward passes, a feature also of back-propagation. However, Qianli Liao et al. (2016) question how necessary this weight symmetry is, and found that:

- The magnitudes of feedback weights do not matter to performance.
- The signs of feedback weights do matter – the more, the better if signs between feedforward and corresponding feedback connections are concordant.

- With feedback weights having random magnitudes and 100% concordant signs, the authors were able to achieve the same or better performance than stochastic gradient descent.
- For asymmetric backpropagation to work some normalizations/stabilizations are indispensable, namely batch normalization and/or "Batch Manhattan" update rule.

The learning method proposed by Sindy Löwe et al. (2020) overcomes many of the limitations of backpropagation, with greater biological plausibility, noting that the brain is highly modular and learns using local information. Instead of a global error signal, a deep neural network is split into a stack of gradient-isolated modules. Training is achieved with self-supervised representation learning (named greedy infoMax algorithm in the paper, see also Oord et al. (2019) so unlabeled datasets can be used, a distinct advantage over the biologically implausible backpropagation which requires labeled data. Each module is trained to maximally preserve the information of its inputs. The algorithm was applied to audio and visual classification tasks with good results. The authors note that as the method enables optimization of separate modules to be performed asynchronously, it is therefore suitable for large-scale distributed training of DLNNs with unlabeled datasets.

8.7.2 Reinforcement learning

The development of reinforcement learning (RL) and in particular the combination of RL and deep learning is a focus at Google DeepMind. A breakthrough in combining these techniques was achieved by Volodymyr Minh and team (2013) applied to Atari game playing. DeepMind has gone on to make multiple breakthroughs with this technology in playing Go, Chess, and other applications. The authors stored past experiences and used these memories compared with new experiences to help drive learning.

RL is a learning method that can be implemented with many different algorithms and architectures (Sutton and Barto, 2018). RL is an unsupervised method (during trials, see below) and conceptually is the closest ML algorithm to how the brain works. There is interaction between a model or agent with its environment, which can be real world physical or a simulation, and the model/agent learns to achieve tasks through interacting with this environment. The brain normally has a goal or intention when performing some tasks and is continually monitoring the success or failure of achieving that goal. The goal in RL can be to maximize a numerical reward signal and the model/agent is not instructed how to achieve the task, but through trial and error it must discover the actions yielding the highest reward. After a sequence of trials, the model is instructed how well it has performed (this is the only unsupervised aspect but is reflective of an intelligence that knows whether the task has been performed or not).

This is the reinforcement signal, and this provides valuable feedback to guide its next sequence of trials.

Matthew Botvinick and the team at DeepMind have explored the implications of deep RL to neuroscience (Botvinick et al., 2020). Looking at the way in which learned information is represented in a deep neural network, the authors examine the processes by which representations support, and are shaped by, reward-driven learning and decision making. The authors see the combination of deep learning and RL each contribute to new emergent patterns of behavior not seen in these methods alone. A drawback with RL is the sparsity of rewards to the number of actions performed in trials (e.g., think of all the moves in a game of chess). There is also a challenge of overfitting and lack of generality in inferencing. Deep RL can overcome these challenges by adding self-supervised learning, where the model/agent produces some auxiliary output to be matched with a naturally available training signal. There is also the challenge that deep RL has a lengthy training process. The long-term temporal credit-assignment challenge remains an area of deep RL research. The use of backpropagation in deep RL is also biologically questionable – see the previous section.

A paper by Chrisantha Fernando and team (2018), also from DeepMind, examined the Baldwin effect in RL algorithms. The Baldwin effect is an alternative to orthodox Darwinian evolution that suggests an organism's ability to survive and adapt in the world and be a successful progenitor of the next generation, has this ability passed on in its genes, thereby passing on these traits to subsequent generations. In learning RL tasks using the Baldwin effect, the authors observed that learning hyperparameters evolved high learning rates. The meta-learning, or learning-to-learn, application of the Balwin effect in RL does not require gradients to be backpropagated to the reference parameters or hyperparameters, and results showed the Baldwin effect capable of producing learning algorithms and models capable of few shot learning in RL tasks.

8.7.3 Natural selection algorithms

Evolution algorithms and computing are part of the standard ML repertoire with wide-ranging applications, principally in solving optimization problems, with many variations such as genetic algorithms, simulated annealing, ant colony optimization, particle swarm, and more (Simon, 2013). Here, we focus on how these algorithms can be related to brain cognition.

Neural Darwinism is an idea proposed by Gerald Edelman in a series of papers and books, see Edelman (1987) and Edelman and Tononi (2000), the gist of which is that natural selection operates in real-time in brain cognition. The inspiration is immunology, the subject of Edelman's Nobel prize, where natural selection was found to operate in how the immune system creates antibodies. Edelman applies a natural selection operation to cognitive function, away from the paradoxical issue of a homunculus that has the "best seat in the house": an executive in the brain that makes top-down decisions (we return to this subject in Section 10.8). Instead, Edelman favors a bottom-up approach where a sea of neurons is awash with signals and a Darwinian process operates to make sense of this data. The cognitive selection mechanism is guided through reinforcement signals originating in the environment.

While the details of Edelman's theory are controversial in neuroscience (for a critique, see Fernando et al., 2012), the basic idea that Darwinian or evolutionary selection processes are operating in brain cognition has been pursued for having potential. In research led by Eörs Szathmáry (Anna Fedor et al., 2017), the team built a cognitive architecture with evolutionary dynamics, they call their model Darwinian neurodynamics (and not to be confused with Edelman's neural Darwinism). The researchers conceive that unconscious problem solving, before conscious awareness of a solution, is a Darwinian process. The model evolves patterns of candidate solutions to a problem, which are then stored and reproduced by a population of attractor networks. The authors point out that what is currently missing in cognitive science is a model that explains how new hypotheses and knowledge are created. The model is seen as a complement to Bayesian cognitive models and builds on earlier work by these authors and others.

To test their model Fedor and team applied it to the four-tree problem, a type of problem requiring insights, developed by Edward de Bono (1967) – famous for his concept of lateral thinking. The problem requires a landscape gardener to plant four trees so that each is the same distance from the other. This type of problem initially leads to an impasse while subconsciously the brain is continuing to work on it, eventually the solution pops up into consciousness. In several computer experiments to solve this problem the authors found Darwinian neurodynamics to be a promising model for how humans solve this type of challenge (see also Czégel et al., 2021).

Biology has been going through a transformation with what is called "the modern synthesis", combining ideas of Darwin and Wallace on evolution and Mendel on inheritance. As evidence mounts that non-genetic inheritance needs to be considered as well, this augmentation of the theory is called "extended evolutionary synthesis". Sizhe Yuen et al. (2023) have investigated how evolutionary computing has adopted these new modes of Darwinism, and as can be seen in Figure 8.13 from their paper, there are ideas in extended evolutionary synthesis that have not been considered in any algorithm, notably epigenetics.

Epigenetics is perhaps one of the most radical ideas to be adopted in recent biology: it is how the environment and an individual's behavior can cause changes in how genes operate, specifically how the DNA is read. Depending on context, DNA, instead of read wholly or not at all, like a light switch

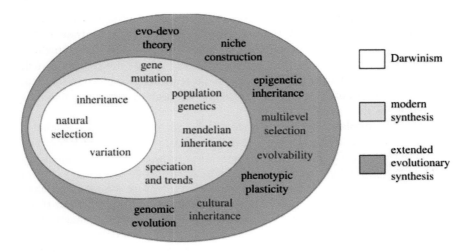

Figure 8.13 Key concepts of Darwinism, the modern synthesis and the extended evolutionary synthesis. Highlighted in green are concepts currently explored in evolutionary computation and swarm intelligence algorithms.

Source: Yuen et al. (2023). With permission.

that is set on or off, is instead set like a dimmer switch (metaphor used in Moore, 2017). Yuen et al. (2023) suggest that epigenetic mechanisms may improve learning rate by spreading changes through a population faster than genetic evolution.

8.7.4 Biologically plausible neural networks

Dorothy Tse et al. (2007) showed in experiments with rats that memory encoding consolidation can occur quickly if an associative schema into which new information is incorporated has been created. The schema concept is associated with systems of neural memory consolidation, i.e., past learning, but experiments were performed to rule out mere familiarity with the environment and task, by sufficiently introducing variation in environments. The results show the rate at which systems consolidation occurs in the neocortex can be influenced by what is already known.

This concept was implemented in a neural network by Tiffany Hwu and Jefferey Krichmar (2020) to study the creation and maintenance of schemas. Their model had two processing streams of indexing and representation and introduced neuromodulation through familiarity and novelty processes which contributed to rapid encoding within consistent schemas. Neuromodulation was implemented as follows: familiarity was modeled by persistent strengthening of weights (long-term potentiation of synapses) in the model's medial prefrontal cortex, with each epoch of training exposing the schema, weight increases equate with familiarity. Novelty is the encountering of new experiences, so in the model's dorsal hippocampus (where weights are initiated with high values) weights undergo long-term depression applied with an anti-Hebbian learning rule when a new stimulus occurs. The authors believe their neural network model is a basis for creating AI with context-dependent memories while avoiding catastrophic forgetting.

One of the implausible aspects of backpropagation is its use of gradient descent and it is difficult to envisage how evolution bootstraps a biological neural network with an implementation of calculus using non-local information and requiring supervised information, i.e., labeled data, to perform the error calculation. This has led researchers (Neal and Hinton, 1998) to explore alternatives that require no calculus, such as the expectation-maximization (EM) algorithm (Dempster et al., 1977) and an unsupervised learning regime. EM is a maximum likelihood algorithm that iterates expectation and maximization steps until the system converges to an estimate of the unknown variable.

Karl Friston has applied the EM algorithm to minimize the thermodynamic concept of Helmholtz free energy, and this approach has led to the free-energy principle for action, perception, and learning in the brain (Friston, 2003; Friston et al., 2006; Friston, 2010). By bringing in this concept of free energy, it is possible to devise a biologically plausible method for learning in the brain. The free-energy principle states that "any self-organizing system that is at equilibrium with its environment must minimize its free energy" (Friston, 2010), i.e., the brain changes to minimize its free energy. The brain minimizes free energy by changing its configuration to affect the way it senses or acts on the environment or change the information it encodes. Free energy is defined as the difference between the probability distribution of environmental quantities that act on the brain and a probability distribution encoded by the brain's configuration. The brain initiates action to minimize prediction errors and perception is used to optimize predictions.

Friston and team (Isomura et al., 2023) have experimentally validated the free-energy principle with *in vitro* rat cortical neurons forming a neural network. The network receiving electrical stimuli causes the neurons to self-organize to selectively encode the sources of stimuli. As predicted by the free-energy principle, changes in effective synaptic connectivity reduced the variational free energy. The authors show that variational free energy minimization can quantitatively predict the self-organization of neuronal networks in terms of their responses and plasticity, and this work validates the applicability of the free-energy principle. See also Section 9.4 for another example of the applicability of the free-energy principle.

We end with a study that looked to build a DLNN that would replicate the behavior of a single human brain neuron (Beniaguev et al., 2021), a cortical pyramidal cell (L5PC) receiving AMPA and GABAA synapses. The

I/O relationship of this physical neuron was fitted onto the artificial neural network, so that the artificial network is trained on identical synaptic input and axonal output of the physical neuron and replicates the I/O behavior of the physical neuron. What is interesting in this study is that the best artificial model, the simplest DLNN that captures the I/O properties of the single real neuron data need seven layers, comprising 128 neurons per layer.

8.7.5 Causal inference

The lack of a systematic and formal causality framework would entail a high degree of subjective conjecture informing cause and effect. To take the example of a barometer (Pearl and Mackenzie, 2018), the expression $B = kP$, where B is the barometer reading, k is a constant, and P is the atmospheric pressure, does not contain information of cause and effect. The equation could be recast as $P = B/k$, but it is our subjective knowledge that the barometer does not cause the atmospheric pressure to change, rather it is the other way around.

Judea Pearl, computer scientist and philosopher, one of the originators of Bayesian networks (Pearl, 1988), has in recent years focused on the question of causality and how to frame it in an objective way (Pearl, 2009; Pearl and Mackenzie, 2018). According to Pearl, the lack of causality formalism has led to the discipline of statistics becoming bogged down with the statement "correlation is not causation" and veering away from questions of causality in favor of simply expressing relations. Pearl and others are working to change this view, with applications in many fields, from clinical pharmaceutical trials to how cloud services retain their subscribers. Pearl (Pearl and Mackenzie, 2018) divides the pursuit of understanding causality into three levels, in ascending degrees of intelligence:

- *Association*: simply observing and seeing how variables are related.
- *Intervention*: Performing an action with the intent of learning about relations and effects.
- *Counterfactuals*: Deeper thinking, such as using the imagination and counterfactual examples to explore, uncover and understand cause and effect.

There are also implications for AI. Pearl places current AI/ML capabilities at the association level. Essentially machines that fit a curve or function to observational data, statistical machines driven by external data, lacking an internal model of reality. Having such an internal model means the AI system may apply volitional forms of learning, i.e., driven from within, and not by an external agent, such as explicitly programmed in.

Intervention is where we change the world through some action designed to discover truths. This is typically performed in the real world, but a good model may also act as a proxy for the world and intervention experiments could be analyzed on it, either theoretically with mathematics or through simulation.

The final level of causation goes one step further, demanding turning time back and asking what if a different choice was made, what would be the consequences, how would it differ from the actual path chosen. Strictly counterfactuals are different from statistics, the latter focused on the available data rather than imagination. However, a good model of the world that is also a function of time is a proxy to the flow of reality in time, allowing running experiments with different choices at different times and generating data that can be statistically analyzed.

An important concept in what can go wrong in causal inferences is confounders. A causal effect can be tested by holding all variables constant and changing only the variable of interest, while running a control group where no change is performed. It is essential that in the test experiment only one variable is changed, if more than one variable is changed, then it is not possible to attribute cause and effect to the variable of interest. Confounders are one or more hidden variables the experimenters are not aware of, but which change between the test and the control experiments. A good design of causal inference of experiments is therefore vital.

There are now available causal inference libraries for use with data science and ML. QuantumBlack, a part of McKinsey & Co., has released open source CausalNex (see https://github.com/mckinsey/causalnex), a library for causal reasoning using Bayesian networks, allowing "what-if" analysis to be performed. A team at IBM has published an open source toolkit (https://github.com/BiomedSciAI/causallib) and accompanying paper (Shimoni et al., 2019). CausaLens (causalens.com) is a company offering commercial solutions in causal inference, including a Causal AI toolbox. Adding causality into AI/ML models is beginning to be explored, see Bernhard Schölkopf (2019; Schölkopf et al., 2021).

8.7.6 Spike-based computation

Simon Thorpe studies possible coding schemes for the brain based on spiking neurons. In Thorpe, Delorme, and Van Rullen (2001), the authors discuss a coding scheme whereby information is encoded in the relative timing of spikes, specifically in the order of which neurons fire. In favor of this scheme the authors cite speed, robustness, and ease of implementation. The authors investigate schemes where a neuron only gets to fire either none or one spike in a 10 ms window, consistent with the fast-processing speed of animal brains. Indeed, the speed with which visual processing occurs places a constraint on the information processing scheme and therefore schemes must be based on just a single feedforward pass through the visual system. Thorpe argues, based on neurophysiological data, that neurons in any given layer have time to generate one or two spikes before neurons in the next layer must respond and enable completion of the processing within about 100 ms.

Taking a population of neurons, in which each neuron can fire or not, a coding scheme is suggested by what the authors call rank order: attributing a rank order number based on the time order of firing. Thorpe (1990) argued that sensory systems and the sensory stimulus itself carry precise (less than a millisecond) timing difference, and this can be used in the coding scheme. There is a physiological process by which relative spike arrival time can be processed by the next layer whereby the first spike to arrive acts as an inhibitor on the next spike. Synchronization of signaling would need re-sets, possible through inhibitory processes, for example, occurring during eye saccades, or the firing of neurons affected by the negative part of local field potential oscillations (see Section 3.16): strongly activated neurons would fire earlier in this cycle. Thorpe shows that relative spike arrival time can be used as an analog code that results in a winner-takes-all processing in the receiving layer, where the strongest and first spike to arrive causes inhibition in the neighboring neurons via lateral interneuron connections.

A key concept discussed by Thorpe, Delorme, and Van Rullen (2001) is to think of neurons as an analog-to-delay converter, see Figure 8.14.

Thorpe and Gautrais (1996) have simulated visual processing using spike order of firing coding in a simulator called SpikeNet, demonstrating sophisticated processing is possible by limiting output cells to one spike per neuron per image. An unsupervised learning algorithm developed at CERCO, Toulouse, France, by Simon Thorpe and colleagues has been licensed to BrainChip, which has created a neuromorphic system-on-chip called Akida, which includes a spiking neuron adaptive processor, implementing the

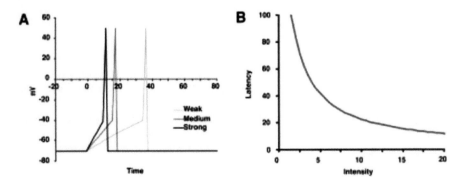

Figure 8.14 (A) Illustration of how a simple integrate-and-fire model of a neuron will produce spikes whose latency depends on the intensity of the stimulation. (B) A typical latency-intensity curve showing how the latency of the first spike will vary as a function of input strength.

Source: Thorpe et al. (2001). With permission.

learning technology. Further development work on related coding schemes has been conducted by Thorpe and colleagues (Bonilla et al., 2022).

In Thorpe (2023), the author sets the case for why floating-point numbers used in DLNNs to represent continuous value firing by neurons are not a good substitute for spikes. Floating-point numbers are used to represent neuron firing rates, the more spikes are fired per second the higher the floating-point number representation. Thorpe gives four reasons why spike-based representations are superior:

1 Floating-point number, continuous value representations of neuron signals are less efficient than spike-based representations, which is why for example, the most energy-efficient neuromorphic chips use spike signals, and indeed the human brain only consumes around 20 watts.
2 Coding schemes that use precise timing of spikes, e.g., see Thorpe (2023), are highly energy efficient.
3 Representing DLNN weights as floating-point numbers is also highly inefficient, modeling neural networks with sparse spiking activity and binary synaptic weights can work effectively and efficiently.
4 Temporal representations of spiking lead to rich information processing dynamics for brain models.

Finally, current neural network models make use of discretized time, and moreover, most neural networks work with neurons that are not functions of time, only of input states, whereas in the real world, time is continuous. Chris Eliasmith and Michael Furlong (2021) propose arguments in favor of learning algorithms that exist in continuous time.

8.8 HYPERDIMENSIONAL COMPUTING

The basic concept in hyperdimensional computing (HDC) and related models such as holographic reduced representation (Plate, 2003), vector symbolic architecture, random indexing, semantic indexing, and SPA (see Section 8.2.3), is that information about objects are represented by very high-dimensional vectors (e.g., 10,000 dimensions) and embedded in a hyperdimensional space. Given the large capacity of such a space, for example, a binary vector would hold $2^{10,000}$ elements, most of the space remains empty and can be used as a sparse distributed memory (SDM: Kanerva, 1990). For an example of the capacity of SDM, see Section 8.5.2.

Taking vision as an example, all the details captured in an image can be represented in a single vector – an ML algorithm can convert a 2D image into a vector with numeric entries that convey all the salient features in the image: the different objects recognized, the type of environment, and the attributes of objects and environment. The spaces being explored are only approximately orthogonal, which leads to an even greater number of possible vectors than in a conventional hyperdimensional space with orthogonal axes.

The hyperdimensional space has useful properties for computation (Kanerva, 2009; Kleyko et al., 2022a, b, Parts 1 and 2); for example, multiplying two vectors combines the information contained in both vectors into a new representation and results in a nearly orthogonal vector. It is also possible to retrieve the individual vectors from the resultant vector. Adding two vectors combines the concepts contained in the individual vectors, and again, the reverse process is possible. The third key operation is permutation, which moves vector elements, say, one element forward and cycles back with the last element. The combination of add and permute operations can represent a sequence of events. Researchers are exploring different vector representations: binary, real, complex, etc.

Pentti Kanerva (2009) provides an overview of the key concepts in HDC. One of the paradoxes of HDC is that where high dimensionality has been considered a "curse" in traditional AI/ML work with means sought to reduce dimensionality of the input space, for example, by using principal component analysis, in the case of HDC high dimensionality is exploited. A positive feature once hyperdimensions are used is the robustness of vectors to noise. When noise corrupts two identical vectors, they will retain their similarity despite a large proportion of the elements differing – allowable errors increase with dimensionality.

To take a binary space as example with vector lengths of 10,000 bits, any vector represents a corner in the hyperspace and its distance to the other corners always follows the same distribution of distances. Typical metrics for distance are Hamming or Euclidean and tend to cluster around the 5,000 bits distance. So, two vectors taken at random will differ in 5,000 bits, and this is true for the whole space. This distribution makes the space robust to noise: more than a third of bits can change between two initially identical vectors and the resulting vectors will still be recognized as related.

A recent success in applying HDC to Raven's progressive matrices (Carpenter et al., 1990), a pattern-based intelligence test, was published by Abbas Rahimi's team (Hersche et al., 2023). In this test a sequence of images is shown containing patterns, and the test is to complete the sequence. In a typical test the HDC system was shown eight options to choose from, one of which was correct. The HDC system was compared with state-of-the-art alternative architectures, including deep neural networks, and HDC had the highest accuracy at 87.7%.

Chapter 9

AI hardware

9.1 INTRODUCTION

This chapter is concerned with hardware, and there is clearly a connection with neurorobotics covered in Chapter 7, but here the focus is on hardware processors designed to emulate cognitive functions. This is the field of neuromorphic computing (Christensen et al., 2022). When the senses of the human body receive external analog stimuli, and this is the case for all senses, receiving light, smell, sound, and touch signals, they process these in local nervous systems and generate electrical spikes (action potentials) that are transmitted to the brain. The brain receives a myriad of these signals and has functional centers that specialize in processing specific senses, but the signals are just spikes. And neurons are interchangeable: when a sense or limb is lost, those neurons that used to process the incoming signals whither and get squeezed out by active neurons. If neurons are damaged in the brain other neurons can take over the function performed by those neurons. Artificial limbs and sensors attached to the body, and which send signal spikes to the brain will be picked up by neurons and sense will be made out of those signals (Eagleman, 2021).

Neuromorphic computing is a branch of computing that is inspired by biology, aiming to be closer to the way the brain works, and this largely means processing with spikes (but note some neuromorphic processors work with continuous value signals).

Note: The selection of processors covered here is a representative sample of leading devices and is not exhaustive.

9.2 NEUROMORPHIC PROCESSORS

The market for companies that produce processors with the neuromorphic label does vary in how the signals are transmitted:

- Spiking neural networks (SNNs) in an analog system, such as electric circuits exploiting Kirchhoff's laws.

DOI: 10.1201/9781003507864-12

	Deep learning/GPU	Neuromorphic chip	Human brain
Power requirements	100-1000 watts	microwatts-milliwatts	12 watts
Processing speed	milliseconds	nanoseconds	milliseconds
Efficiency	Variable	High	High
Learning rule	Global	Local	Local
Event-based processing	Less suitable	Yes	Yes

Figure 9.1 Comparing the brain, neuromorphic chip, and GPU in AI inference mode.
Source: E. M. Azoff.

- Digital processors implementing SNN.
- Analog device with non-spiking, continuous value signal neural networks.

Neuromorphic engineering or computing aims to be biologically plausible, recognizing that biological (i.e., analog) systems are far more efficient than digital systems. A common characteristic of neuromorphic processors is that they have low power consumption, high efficiency (typically in the form of highly sparse connectivity) – both characteristics of the human brain – and fast processing speed, see Figure 9.1.

Neuromorphic processors on analog technology are primarily used for AI inferencing, and use integer precision or the analog equivalent, and do not support floating point numbers. They are event-driven and so naturally process data in no batch mode, i.e., batch of one.

There is a challenge of training neuromorphic neural networks without the use of a global learning rule, such as backpropagation. Backpropagation is critical in non-spiking DLNNs, where it uses information at the output of the network to update synapse weights upstream in the network. There is no evidence for any such global learning rule in the human brain and neuromorphic computing research has led to the development of local learning rules.

As with traditional processors for AI, there is one market for training neuromorphic artificial neural networks and a separate market for inferencing neuromorphic artificial neural networks. The key differences are:

- *Training*:
  - Backpropagation (for global learning algorithms) requires a forward and backward cycle, and this is typically repeated millions of times for large DLNNs.
  - A calculation performed repeatedly in backpropagation is multiply and accumulation operations that run well on GPUs.
  - Calculation requirements for training neuromorphic architecture algorithms varies with the producer of the processor.

- *Inferencing*:
  - Once a trained artificial neural network is achieved the inference mode requires only a single forward pass at each presentation of an input pattern. The calculation performed for an inference is relatively modest compared to training mode, but inferencing is typically performed many times: for millions of input processing, the need for efficiency is paramount.
  - Event-based processing in real time: a key difference between traditional neural networks and neuromorphic neural networks is that the latter is designed to process signals in real time, responding to events as they happen.

Strategies for training neuromorphic neural networks include using spike timing-dependent plasticity (STDP) rules (Andrade-Talavera et al., 2023; Maass and Schmitt, 1999), adapting backpropagation and backpropagation through time to train SNNs, and transferring DLNNs trained on a CPU and GPU combination onto a neuromorphic SNN architecture. Research into local training rules in most recent years led to energy principles (see Section 8.7.4) to train SNNs on neuromorphic architectures, for example, equilibrium propagation (Ernoult et al., 2020) and eligibility propagation (Bellec et al., 2020).

Most neuromorphic processors use standard CMOS technology on legacy semiconductor wafer fabrication nodes, which makes them highly cost effective, and reliable, based on proven technology. The neuromorphic processor is also highly energy efficient due to exploiting of two forms of sparsity. The first, activation sparsity, has to do with sparsity in the sensor activation which becomes the SNN input data. For example, with vision as the sensor reacts to events, pixels only fire when a change is detected (above a preset threshold), such as an object moving in the field of vision. Operating in real time, as pixels fire they activate a small subset of neurons, which is highly energy efficient; if the object view is static, then no neurons are fired up, very little energy is consumed, and the device can be left always on. The second type of sparsity is in the processing, the network works asynchronously, with only a few neurons firing up across the sparsely connected network. This leads to rapid and energy-efficient processing. This sparsity explains why digital SNNs can be as energy efficient as analog SNNs. Edge devices powered by battery can last in the field for 12–18 months, which could not be done with a GPU.

Latency in neuromorphic processors must be measured in the context of the application: the technical latency is nanoseconds, but to gain a high performance out of the network, input signals need to be integrated over a period, e.g., a single vision event doesn't yield much information, so it's the application latency that needs to be quantified.

Two companies building neuromorphic chips are Intel and Rain Neuromorphics, examined next. See also Brain Chip's Akida neuromorphic processor in Section 8.7.6.

9.2.1 Intel's neuromorphic computing research

Intel launched its Neuromorphic Research Community (NRC) program in 2018 along with its first-generation neuromorphic chip, Loihi, available to research labs for exploring the potential of neuromorphic computing (Davies et al., 2021). In 2021, they launched Loihi 2, based on fabrication process node Intel 4, it has 2.3 billion transistors, a maximum of 1 million neurons per chip with 120 million synapses per chip. The neuron model is fully programmable using the open source Intel Lava software. This greater flexibility allows variations of the backpropagation algorithm suitable for SNN to be supported. Spike information can be coded in up to 32-bit payloads. Loihi 2 supports minimum chip-wide time steps under 200 ns allowing neuromorphic networks to process up to 5,000 times faster than biological neurons.

In the time this research program has been running, Intel NRC members have evaluated Loihi in a wide range of application demonstrations, including:

- Adaptive robot arm control.
- Visual-tactile sensory perception.
- Learning and recognizing new odors and gestures.
- Drone motor control with state-of-the-art latency in response to visual input.
- Fast database similarity search.
- Modeling diffusion processes for scientific computing applications.
- Solving hard optimization problems such as railway scheduling.

In these applications Loihi typically consumes far less than 1 W of power, demonstrating breakthroughs in energy efficiency.

9.2.2 Rain neuromorphics

Rain Neuromorphics is a startup founded in 2017 and based in Silicon Valley. Its focus is on building analog processors incorporating neuromorphic principles, and its long-term aim is to make its AI hardware match the scale and efficiency of the human brain, with its 86 billion neurons, 500 trillion synapses, and operating at around 12 W. The venture is building an analog processing unit (APU), its estimated power consumption will be 0.1% of the Nvidia A100 GPU and running between 7k to 20k faster.

The APU has physical equivalents for neurons and synapses based on a compute in-memory resistive RAM (ReRAM), also known as memristor. In the Rain architecture the use of digital to analog converters (DAC) and analog to digital converters (ADC) are necessary only once when data is input and output respectively to the chip, resulting in significant power savings. Rain believes this aspect alone will reduce power by 90% and chip space by 75%. Rain describes APU as the first fully physical analog neural network and overcomes the von Neumann architecture bottleneck.

In the APU there is no separate weight storage, the weights in the analog domain are stored at the synapses in the ReRAM elements. There may be some off-chip processing for pre-processing data before being streamed into the neural network, but the neural network is entirely a non-von Neumann architecture. In some sense, the architecture can also be described as a data flow architecture, where data flows through the architecture without data having to be fetched to and from a neural core by some digital clock beat.

Rain identifies three key challenges working with analog neural networks

- The initial signals can be noisy and the noise compounds as signals travel through the layers.
- The same inputs can result in different outcomes.
- Natural device-to-device variation can add inconsistencies.

Given that animal brains overcome these limitations gives motivation to find an analog brain algorithm that is similarly robust. Rain uses equilibrium propagation developed by Yoshua Bengio and his team (Ernoult et al., 2020) who had analog learning systems in mind when designing the algorithm. It is an energy-based model for training analog neural networks and is able to exploit the local gradient information available. This led Rain co-founder Jack Kendall and Bengio to develop the proprietary algorithm used by Rain.

Equilibrium propagation has the power of backpropagation while also satisfying local information constraints. It gains its gradient information through energy minimization using a local learning rule. This involves measuring the voltage differential at any synapse and working out whether to raise or lower its weight at that point to learn the information present. The original algorithm from Bengio's team effort related backpropagation to energy-based methods. In 2020, Rain and Bengio further developed this algorithm to apply to the APU's analog circuit (Kendall et al., 2020). The design of the circuit allows the algorithm to be derived directly from Kirchhoff's laws. Being energy based makes it ideal for analog systems where the energy is a physical quantity, in Rain's case power, that is present and measurable.

Rain describes inference in the APU as running a wave of electricity through the physical network: an electric circuit. As the electricity flows, obeying Kirchhoff's laws, it performs the equivalent of matrix mathematics.

The algorithm Rain uses has been patented and in its first version is designed for static input circuits, and Rain already has a patent application in the works for dynamic input circuits, which extends the original equilibrium propagation algorithm.

9.3 NEURRAM ANALOG CHIP

Researchers from Stanford University including Weier Wan, Philip Wong, and Gert Cauwenberghs (Wan et al., 2022) have developed an analog chip that performs compute in memory and exploits a new type of non-volatile memory called resistive RAM (RRAM), of which there are 48 cores, each containing 256×256 RRAM cells and 256 CMOS neuron circuits that perform analog to digital conversion and implement activation functions; the new chip is called NeuRRAM and built on 130 nm CMOS and RRAM foundry process technologies. The first example of the chip has 3 million memory RRAM cells in total. The new chip was designed to simultaneously co-optimize three dimensions of the device: energy efficiency, computational versatility, and numerical accuracy. The chip is claimed to be 1,000 times more energy efficient than digital equivalents. The NeuRRAM architecture is a dynamic reconfigurable dataflow which is highly versatile.

The authors find that NeuRRAM out-performs digital equivalents, its inference accuracy is comparable to software models quantized to four-bit weights, and its performance on standard AI benchmarks includes 99.0% accuracy on MNIST18, 85.7% on CIFAR-1019 image classification, and 84.7% accuracy on Google speech command recognition.

The team expects further development on NeuRRAM such as moving to 7 nm technologies will yield peak energy efficiency improvements by two to three orders of magnitude while supporting bigger AI models.

9.4 NVIDIA AI GPUS

Nvidia, GPU chip manufacturer, has year-on-year improved the performance of what was once a GPU designed for playing video games and turned it into one of the world's leading AI accelerator chips. The GPU can efficiently accelerate the "multiply and accumulate" operations required to train DLNN models requiring millions of iterations in large models.

Historically, we saw in Part Zero how the GPGPU, first introduced by Nvidia, accelerated DLNN training times from weeks and months (depending on the model size) to hours and days, and enabled the deep learning revolution. Today top-end AI-targeted GPUs play a role in accelerating the training of the largest generative AI and LLM models (see Section 8.6.2) with many billions of parameters. Smaller DLNN models may use alternate training accelerators, such as based on advanced CPUs, software acceleration methods, field programmable gate arrays, custom acceleration chips, or some combination of these methods and devices.

9.5 IN VITRO NEURONS LEARN TO PLAY PONG

Researchers at Cortical Labs and several academic institutions (Kagan, 2022) created in vitro neurons (live neurons in a dish) and taught them to play Pong, an early one-player Atari computer game: they named the system DishBrain. In Pong, the player controls a "bat" moving in a vertical line on the left "wall" to meet a "ball", a moving spot on the screen, which obeys Newton's laws of motion and bounces off all the walls of the square court space or the bat if the two make contact.

While the game is a video simulation, the neurons playing it are quite real. Brett Kagan (2022) created several test cultures, each with about 800,000 neurons, and formed as a single layer on a high-density multielectrode array base that could both stimulate the neurons with action potentials like natural ones, as well as read electrical activity. Some dishes comprised neuron cultures sourced from embryonic mice, and other dishes comprised neuron cultures sourced from human stem cells. Cultures created in this way could be kept alive for months. The authors report that over time the neurons formed numerous and dense dendritic and axonal connections with supporting glial cells. The human stem cells started forming a network after about 30 days, this growth was confirmed via scanning electron microscopy on cultures maintained for three months. The authors name these cultures biological neural networks (BNNs). The human cells first started showing electrophysiological properties (synchronized spike bursting) at 73 days in vitro, while mouse cells started showing these properties after 14 days in vitro.

DishBrain was operated as follows. There was no concept of reward or punishment to drive learning in the experiments conducted, instead, a BNN was connected to the video game via electrical stimulus that depended on the game status. The position of the ball on the screen was translated into a steady stream of stimuli (sensory information) across the multielectrode array. An area of the multielectrode array was designated as the "motor" region and its outputs were used to move the bat. The following feedback systems were developed using custom software drivers and eight electrodes activated at fixed points in the multielectrode array:

- Closed-loop feedback where the bat is moved by BNN signals, during game progression before the ball reaches the left wall, the electrodes are sent regular sensory stimulation

  - *Stimulus state A*: If the ball made a hit on the left "home" wall, i.e., the bat missed the ball, then a spike of a spike of 150 mV voltage at 5 Hz for four seconds was sent to a randomly selected electrode in the eighth.
  - *Stimulus state B*: If the bat made contact with the ball, then a predictable spike at 100 Hz for 100 ms was sent to the BNN, stimulating all eight electrodes.

- *Stimulus state C*: In silent feedback, whether the ball hits the home wall or meets the bat, no result stimulus is transmitted, only ball position sensory information.

- *Stimulus state D*: Open-loop feedback, the bat is not connected to the BNN so the bat is stationary in the middle of the home wall, and ball position is tracked through stimuli to the BNN. There are no stimuli transmitted when the ball hits the home wall or the bat.
- *Stimulus state E*: Rest position: there is no signal transmitted from the video game to the BNN and no signal from the BNN to the video game.

For a system to demonstrate intelligence it must possess two processes, namely perception, where external states influence internal states, and action, where internal states influence external states, and combine these with learning how its actions can influence the external environment using its perception. Of course, to the BNN, there is no understanding of "internal" and "external", it exists within its own world.

As Brett Kagan (2022) explain, Friston's free-energy principle (see Section 8.7.4) in a system like DishBrain will see it seek to minimize its free energy, which allows a test of the principle. The gap or error between the model predictions and observed sensations may be minimized by aligning predictions with sensations, or by acting on the environment to make sensations conform with predictions. According to this notion, BNN behavior can be shaped by sending unpredictable feedback, stimulus state A above, following incorrect behavior, i.e., bat misses ball. The theory states that the BNN will adopt action that avoids states that result in unpredictable input. The question posed by Kagan is thus whether a BNN like DishBrain can engage in goal-directed learning, at the same time testing the free-energy principle.

Control cultures were created with BNNs that were not stimulated by feedback from the game. The designated motor region in the culture was also changed in experiments to ensure there was no bias from one region. The results with the highest hit/miss ratios were obtained for the human cell cultures. Experiments were conducted with increasing rate of stimuli positioning the ball. This improved the length of rallies.

To test the free energy principle, experiments were conducted across the range of states defined above: stimulus state A and B versus stimulus states C to E in control cultures. Only the mice and human non-control cultures showed learning, and interestingly the human non-control BNN, which ultimately showed the best performance, was initially inferior to both mouse and control BNNs.

In conclusion, Brett Kagan applied implications from the free energy principle, and in a series of game experiments found apparent learning within five minutes of real-time gameplay, which was not observed in control conditions. The cultures displayed the ability to self-organize activity in a goal-directed manner in response to sparse sensory information about the consequences of their actions. The authors call this synthetic biological intelligence.

Cortical Labs (corticallabs.com), based in Melbourne, Australia, was a key participant in this study and is working toward placing DishBrain on the cloud for researchers to access and run experiments.

Part Three

Speculations toward human-level AI

INTRODUCTION

Part One presented a survey of latest findings in neuroscience and Part Two a survey of recent research in AI, selections based on being relevant to the book's mission of building an HLAI system. Having up to now avoided giving my opinions and analysis, this part is mostly about my views and speculations on directions to achieve HLAI, while drawing on Parts One and Two. My intention is to provoke debate and experiment, to stimulate thinking and help move AI research toward achieving HLAI.

Leon Cooper made essential contributions to understanding superconductivity for which he earned a physics Nobel prize winner. This was in the face of many physicists discouraging him from researching the topic for being too intractable. Cooper subsequently turned his research focus to AI. In talking about his research career, he shared his three guiding precepts for research to James Anderson and Edward Rosenfeld (2000).

1 Don't attack a complicated problem if there is a much simpler version that you are unable to solve. Solve the simple one first.
2 Don't believe what you don't understand.
3 Beware of those who say that the solution is in the complexity – that there is no way to see what's going on. Of course, in some cases this may be true, but most of the time it's just a way of throwing in the towel.

Given the complexity of the human brain and the lack of a theory to explain how it works, the mission to crack that problem can well heed these precepts.

This book puts forward the proposition that a computer simulation can take us toward HLAI. One notable impression from reading Dennis Bray's Wetware (2009) is the sophistication of the most elementary life form, the simple prokaryote cell. It has no nucleus, but its most important feature is the cell wall which separates the soup of chemicals and other life forms that exist outside the wall, and the soup of chemicals inside the cell. This lifeform has no brain and no consciousness, the best way to describe it is as

DOI: 10.1201/9781003507864-13

a self-sustaining and reproducing equivalent of a computer simulation: run not by electricity but by chemicals and their reactions, with diffusion and random motion propelling the chemicals inside the cell. This simple cell is the living embodiment of a computer program and given this, the stretch to a computer simulation of HLAI appears to me now less strange.

Finally, the scientific method will be raised here in two ways: one, that this method is the means to discover the workings of the brain (this may seem an odd statement to some, but there are some people who say the brain has some quality beyond our capability to fathom it). Second, that the scientific method, as a process, should be built into an HLAI system. For completeness, I have left to the Appendix to say a few words on what I mean by the scientific method.

Chapter 10

The possibility of creating HLAI

10.1 THREE TYPES OF HLAI

There are three approaches to building HLAI. The first is to understand how the human brain works and with this knowledge build a human-like human-level AI, or $(HL)^2AI$. This approach takes its inspiration from neuroscience. Our brain is the ultimate example to which we aspire to build an intelligent machine: humans have the highest form of intelligence we know. This motivates the study of neuroscience to guide building the $(HL)^2AI$ machine.

The second approach is to build an intelligent machine on ideas rooted in engineering (knowledge engineering, computer science, mathematics, etc.), let's call this engHLAI. Modern electronic computing including memory technology, long-term storage, and rapid numerical computation is vastly superior in capability to the equivalent features in the human brain, so the modern information technology stack can be leveraged in building capabilities in engHLAI. Many aspects of the human brain at the molecular level are a means to achieve some functional step from A to B. If electronics can shortcut the molecular complexity that occurs in a living body, then using it makes sense. The current research direction of generative AI and LLM (Section 8.6.2) may yet succeed in building HLAI and it would in such a case be an engHLAI – resembling nothing like the human brain but a direction taken in a purely engineering one. However, it is the hypothesis of this book that to first achieve HLAI, taking a pure engHLAI approach is less likely to succeed than taking a $(HL)^2AI$ or hybrid approach to be discussed below.

There is another aspect to engHLAI: exploiting the concepts of AI cognitive architectures (see Section 8.2). Progress with cognitive models has been modest in comparison with recent successes of DLNNs; however, cognitive architectures could play a role in engHLAI models. Adding pre-built information processing structures in the form of cognitive architectures to engHLAI, combined with neural networks that learn about their environment and learn how to solve challenges – this can be compared to how the human genome includes a blueprint for the design

DOI: 10.1201/9781003507864-14

of the brain, and once that design has been instantiated experience takes over in adapting it for use in the real world. Similarly, one can conceive of an engHLAI that uses cognitive architecture to kick-start the AI with designed components that can work in combination with neural networks (Zador, 2019).

The third approach is a hybrid blend of the first and second approaches, hybridHLAI or H²LAI. A hybrid approach exploits the best of computer engineered HLAI combined with human-like HLAI. This may be the ideal approach, exploiting the patterns and concepts we discovered in the human brain and accelerating using engineering algorithms that have no correspondence in the brain but can accelerate computation and shortcut the complexity that exists in in wetware through hardware and software.

Components that we see as useful for HLAI such as LLM (Section 8.6.2), causal inference (Section 8.7.5), and more, can be pre-built and ready to use. There is debate in biology between nature versus nurture. Many animals are born with capabilities that are hard-wired and usable straight from birth, while a creature that can learn is more adaptable to survival in a changeable environment. A blend of the two approaches, pre-built features and adaptable, learning features, may be optimal for an HLAI system.

In Section 11.2, the outline of an H²LAI system is presented, but aiming to build such a system is only one approach. There are two alternatives that can offer important lessons for building HLAI: one approach to crack the hidden neuronal code is by focusing on emulating lesser creatures that show intelligence, with a relatively manageable number of neurons in their brains, the idea is to identify principles that can then be scaled up to emulate HLAI. This would be an animal-level AI (ALAI) – whereby animal we mean not including humans.

Another approach is to build an environment in which evolutionary algorithms can drive learning, and these algorithms can become part of the focus of the learning process, by evolving optimal evolutionary algorithms. The best resultant algorithms can then be embedded in an HLAI system. The process would start with simulations of virtual worlds in which the rudiments of intelligent systems are evolved, fast forwarding evolution of a population of virtual creatures with brains based on ALAI models. The rules of survival and reproduction can be based on Darwinian and neo-Darwinian rules of natural selection.

Once creation of an HLAI system has been achieved, once the functioning of the brain is fully understood, the future beyond (HL)²AI will likely be overtaken by engHLAI, as engineered systems can be further developed. Building subsequent generations of HLAI may be undertaken by intelligent machines and it is then no longer a question of humans developing HLAI but intelligent machines surpassing our creations with transcendent AI.

There is justifiable concern that intelligent machines may, knowingly or unwittingly, harm or destroy humanity. Until we have more confidence in the machines, we build, we should ensure the following two points are always adhered to: make sure humans have sole control of the off switch. And build animal or human-level AI (A/HLAI) systems with behavior safety rules implanted. In Section 11.2, this is described as "preset directives", essentially the developer implants rules that ensure the A/HLAI system does not harm humans.

The flying metaphor has been used to describe the quest for intelligence: man wanted to emulate the birds and so strapped wing contraptions that would flap by human strength or bicycle power, only to fail. But now we have supersonic jet planes that cross continents in a few hours and rockets that take us to the Moon, feats no bird can achieve. The parallel argument for AI is that an engineering approach can take us beyond the constraints of evolutionary creation, and achieve HLAI by other, superior means. This might happen.

However, *understanding* how birds fly, by virtue of a wing's aero foil concept led to the successful first flight of an airplane. From that point onwards engineering led to advances beyond birds. In my view we need to pursue a course of research to understand how the animal or human brain works. Once we have grasped the basic mechanism, the equivalent of the aero foil, we can then apply engineering that takes designs beyond evolution. For a start engHLAI would not need to be constrained by the slower ion-based electrical transmission in the brain, and instead exploit modern high-speed electronics. We could integrate the most advanced computers with an HLAI. But this research path starts with first understanding how the brain works – working from simpler animal brains to the human brain – this may be the quickest path to achieve HLAI: recall Leon Cooper's first precept to solve the simpler unsolved problem before the more complex one.

10.2 NEUROSCIENCE INSPIRATION

One of the challenges in sifting through neuroscience research is to gain a sense of what matters for computation in the sense of being necessary to replicate with engineering and what is just how nature achieves steps and processes with biology and chemicals and not relevant to the process of computation and thinking. This is where the power of abstractions comes in. Modern computing high technology has layer upon layer upon layer of abstractions, so that anyone working with this technology operates within the layer of abstraction of their concern and shielded from irrelevant material. This is badly missing in the mission to re-create the brain's thinking – there are few abstractions and models to help focus the task.

One of the challenges in aligning neuroscience and AI research is that the former has multiple goals, a significant one being to understand healthy operations in the brain to help diagnose, understand, and ultimately cure diseases and ailments of the brain, as well as understanding the biological mechanisms that create intelligence, which aligns with AI. The health implications of neuroscience are major drivers for receiving research funding. However, given the recent successes of DLNN there is a noticeable increase of attention by neuroscientists to the question of the biology of intelligence. To say it is a "holy grail" of neuroscience is not an exaggeration, but it was dismissed until recently for being too complex a challenge, the change is due to improved non-destructive probes and measuring instruments better able to help observe the workings of the brain, yielding more data on which to build models and theories (see Appendix).

In the AI research community, more AI researchers are calling for greater attention to neuroscience to help drive their research (e.g., Hassabis et al., 2017; Sinz et al., 2019; Ramezanian-Panahi et al., 2022). In the extended debate in Brenden Lake et al. (2017) the authors make useful suggestions, such as creating systems that learn to learn. Zador et al. (2023) propose a neuroAI grand challenge: the embodied Turing test. The Turing test is designed to see if we can distinguish between a machine and a human responding to a natural language conversation (the UK's Royal Institution 2023 Christmas lecture featured a test where the audience had to distinguish between OpenAI's ChatGPT and a human behind a screen). The proposed embodied Turing test concerns a neurorobot of some form with advanced sensorimotor abilities interacting with the physical world. The embodied Turing test is to distinguish between the robot and an animal it imitates (including humans, the ultimate test).

An intriguing experiment was conducted by Eric Jonas and Konrad Kording (2017) to see if the data analysis methods used in neuroscience applied to a classical microprocessor, treated as a model organism, could reveal an understanding of the workings of the device. This study found "interesting structure in the data" but did not meaningfully describe the hierarchy of information processing in the microprocessor. This finding reinforces the need to build theories and models of cognition upon which to test ideas against data. Without such models we are in a position not dissimilar from a social anthropologist observing the lights switching on and off in an office building and using this data to figure out what activity takes place in the building.

Christopher Parisien et al. (2008) point out that typical neural network models, such as deep learning, lack biological plausibility in respect of not reflecting that some neurons are excitatory, and some are inhibitory but not both. Certainly, from an engineering perspective, we can build mechanisms that allow artificial neurons to perform both functions, but the point is that current learning algorithms only model excitatory neurons. The role of inhibitory neurons is important as it reduces noise in the brain and allows the strongest and fastest neuron responders to stimuli to influence the brain's thinking process.

10.3 CONSCIOUSNESS AND HLAI

When discussing consciousness, we need to distinguish between levels of conscious thought: principally whether it includes self-awareness (conscious of being conscious) or not – Anil Seth's label for this is "metacognition" (Seth, 2021). In the former when a human is totally absorbed and lose themself in conscious thought, we say the person is "in the flow" (ITF) or "in the zone". When we are deeply concentrated on some work, absorbed reading a novel or engrossed in watching a film, we lose our self-awareness. I speculate most animals are in that conscious "in the flow" state with varying degrees of conscious thought capability (see also Section 4.4), but maybe only higher animals have a meta-conscious level, an awareness of their own consciousness, a sense of self. Hereon, my reference to ITF consciousness (ITFC) will mean consciousness without self-awareness.

I speculate that HLAI is possible without needing to be concerned about conjuring up fully aware consciousness, and that ITFC is an emergent property of visual processing (more on this below). But to address these ideas, let us return to the three questions posed in the Preface:

1 Is an algorithmic simulation of a brain sufficient to make it intelligent?
2 Do you need consciousness to have intelligence?
3 Do you need to be alive to have consciousness?

Here are my views in answer to these questions, in reverse order:

- *Do you need to be alive to have consciousness?* We still do not know what consciousness is, but I speculate it is an emergent property of a brain. To answer the question, we only have evidence for the various forms and levels of consciousness in living creatures, with ITFC likely to be common in many animals. I speculate that ITFC can be emulated in a computer simulation, so we may be able to create it without needing a living system.
- *Do you need consciousness to have intelligence?* In my view, yes. Earlier I said ITFC may be emulated in a computer simulation. This level of consciousness assists an animal in planning and thinking beyond current data, using imagination to foresee alternative futures, or recall relevant incidents from the past and apply lessons learned to the current situation.
- *Is an algorithmic simulation of a brain sufficient to make it intelligent?* In my view, yes, but this is an algorithm that once you switch it on takes on virtual life. If we succeed in building an HLAI system, it will most likely succeed because it possesses an ITFC capability. I speculate that ITFC will emerge from modeling visual processing, leading to visual thinking.

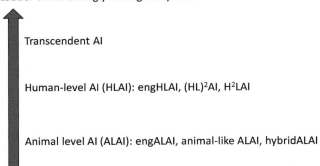

Ladder of increasingly intelligent systems

Transcendent AI

Human-level AI (HLAI): engHLAI, (HL)^{2}AI, H^{2}LAI

Animal level AI (ALAI): engALAI, animal-like ALAI, hybridALAI

Figure 10.1 Levels of increasingly intelligent systems.
Source: E. M. Azoff.

While modeling ITFC may take us toward a *near* HLAI system with our current knowledge, i.e., to a lesser animal-like level, achieving an (HL)^{2}AI system is a further leap in capability which I believe is beyond our current knowledge. Humans are self-aware and this surely can only be explored once we have developed an ALAI system with ITFC (my use of "animal" is meant lesser than human). Figure 10.1 puts together these different systems in terms of levels of intelligence.

10.4 VISUAL THINKING AND CONSCIOUSNESS

> I very rarely think in words at all. A thought comes, and I may try to express it in words afterwards.
>
> Albert Einstein (1959)

First, to address the role of language in HLAI research. LLM (see Section 8.6) offers a first-class language interface with an AI system, but the question is whether to develop an HLAI that includes language skills. My view is that following the principle of walking before running, the focus should be first on visual thinking – some aspects of this debate are presented in Figure 10.2. As expressed above, an ALAI should be a first step toward HLAI and this means solving the visual thinking challenge first.

The question for AI researchers is whether it is easier to create a thinking process based on visual thinking than one based on language, to help decode how neurons communicate information and how the brain uses that information to think. We all evolved from the most primitive life forms so visual thinking must be the first brain thinking process before brains started processing sound for language or singing/music.

	Verbal thinking	Visual thinking
Pros	• LLM offers successful statistical model albeit without understanding. • Human subjects can describe their experiences in research experiments.	• Visual pattern identification of danger, food, and reproduction can be hard wired. • Whole animal kingdom as subjects. Pick a brain with less neurons than humans to study.
Cons	• Challenge of language understanding. • Only have humans as subjects to study, and their brains are the most complex.	• Challenge of vision understanding.

Figure 10.2 Verbal thinking versus visual thinking.

Source: E. M. Azoff.

Verbal consciousness, whether internal dialogue or dialogue with other people, is a skill that developed in human evolution after visual thinking. Language takes consciousness to a higher level and is distinct from sensory consciousness which we share with much of the rest of the animal kingdom. Amit et al. (2017) performed behavioral and fMRI experiments to test verbal and visual thinking and found an asymmetry: a high degree of visual imagery was engaged whether the experiments were verbal in nature or visual, whereas inner speech was largely performed only in the verbal experiments. The conclusion is that visual imagery is a more fundamental property of the human brain than language.

According to William G. Allyn: "More than 50 percent of the cortex, the surface of the brain, is devoted to processing visual information. Understanding how vision works may be a key to understanding how the brain as a whole works" (Hagen, 2012).

Given recent research expressing the view that most animals with a brain, however limited, have a degree of consciousness (not necessarily self-aware consciousness) and that these same animals have visual systems with a large part of the brain devoted to processing vision, I speculate that there is a connection between consciousness and visual thinking (see Sections 4.5 and 4.6): visual thinking is an important part of consciousness (even people blind from birth experience a form of visual thinking, just not fed by external visual inputs).

Visual thinking could also be the key to unlocking the mystery of what is consciousness. Consciousness gives access to planning at a high level, the processing to create the world view in our head is hidden in the subconscious, consciousness allows us to work with the higher abstractions in terms of logical thinking. This is what the visual system does when it takes the raw signals from the eyes and converts them into the 3D world view in our head. Michael Land in his book "Eyes to See" (2018) surveys the many different evolutions of visual systems in the animal kingdom and in a penultimate chapter he speculates that vision is tied up with consciousness. Land speculates that the kind of planning that, for example, jumping spiders (they have four pair of eyes) undertake to ambush a prey while avoiding threats requires a high degree of planning and understanding of the environment one would associate with a conscious being.

This quote by Rudolf Arnheim from his book *Visual Thinking* (1969) sums it up well:

> My contention is that the cognitive operations called thinking are not the privilege of mental processes above and beyond perception but the essential ingredients of perception itself. I am referring to such operations as active exploration, selection, grasping of essentials, simplification, abstraction, analysis and synthesis, completion, correction, comparison, problem solving, as well as combining, separating, putting in context. These operations are not the prerogative of any one mental function: they are the manner in which the minds of both man and animal treat cognitive material at any level. There is no basic difference in this respect between what happens when a person looks at the world directly and when he sits with his eyes closed and 'thinks'.

It seems likely that neurons preoccupied with understanding the environment, movement within the environment, and representing the environment internally is a major function for the brain. Then other neurons can abstract these processes to provide a layer of reasoning about the environment. Brains with neurons that can do that have an evolutionary advantage.

David Eagleman (2021) observes that the way neurons fire (the pattern of neural "storm") when the brain is thinking has striking similarity to when we perform a motor movement, such as moving an arm. Eagleman sees a connection: our arms move objects in physical space, and thinking moves concepts in thought space.

There is a connection between the experience of vision, seeing the external world represented internally in our brain, with the concept of visual consciousness. Working on the assumption that consciousness contributes functions to different psychological domains, such as vision, emotion, and social cognition, has given rise to distinct research avenues, visual consciousness being a major one (Ludwig, 2023). It's my impression these are all just facets of the same phenomena we call consciousness.

Taking the route of ALAI before HLAI, then I make the leap to suggest that visual thinking in animals is equivalent to ITFC. This is modeling an animal brain that can construct a 3D view of the external world, its environment, and apply logical thinking to the objects in that environment, understanding how these objects relate to its prime concerns: threats and dangers, sources of food, shelter from the elements, and reproduction (important for animals but redundant for AI machines).

Furthermore, the thinking part of this skill must be a process of self-learning. For this I propose a simplified embodiment of the scientific method: this is the plan-do-check-act (PDCA) cycle. PDCA is discussed in Section 11.2, for now, I suggest that the process of learning through trial-and-error experiments (in the real world or purely thought experiments), guided by memory, which the PDCA cycle entails, provides the internal thinking drive, which combined with the visual system in ALAI may give rise to ITFC.

Speculating further, the habituation of visual thinking in more evolved animals may give rise to neurons that have learned to abstract that process and lead to fully aware consciousness (and what psychologists call theory of mind – recognizing other individuals as having consciousness like ours). But that is a step of speculation that should be parked for return later, after ALAI is achieved.

Finally, to address views that are expressed in several papers and books on the topic of HLAI and consciousness, that it is a futile endeavor, I will give an example of one paper. Thus, Alfred Gierer (2008) suggests that the problem of what is consciousness has parallels with mathematical theorems of undecidability and that brain analysis cannot surmount the self-referential nature of understanding consciousness. Gierer makes statements such as "the brain-mind relation may not be fully decodable in principle". In my view, it is defeatist, an example of Leon Cooper's "throwing in the towel" in the precepts quoted at the top of this part. I think the scientific method has more to offer on this challenge, but the scientific method is also reliant on our measuring instruments and experimental tools – the method and the tools have always enabled progress through their combination. Our current impasse in unraveling the mystery of the brain and consciousness is not proof of the impossibility of cracking it, and I have no reason to believe the scientific method has its limits: the basic approach has not changed between 300 BC and today, but we have clearly made a lot of progress (see Appendix).

Furthermore, when Gierer and others talk about consciousness, it is the full-blown affair of the human brain, but as we have seen in Section 4.4, the animal kingdom is now understood to possess degrees of consciousness starting with humble insects and going up the intelligence scale. There are different kinds of consciousness and I consider the feature of self-awareness as a significant differentiator between lesser forms of consciousness and those animals that possess self-awareness. Understanding how the brain works, decoding the neural code, is an essential first step before any consideration of visual thinking and beyond. Good models of the brain and simulations of these models could yet break through the barrier of Gierer's unknowability.

The past two decades have also seen a revolution in non-destructive monitoring and observation of the inner workings of the brain, yielding clues on how best to construct our brain models. Innovations like Dishbrain that can grow neurons in the laboratory and make them learn tasks (see Section 9.4) give further confidence that we can discover vital clues to understand the workings of the brain.

10.5 BRAIN LATERALIZATION, VISUAL PROCESSING, AND INTELLIGENCE

The purpose of this section is not to list the many differences between the two hemispheres and make suggestions as to how this impacts human behavior and society (for that see Iain McGilchrist, 2019, 2022). Rather, it is noting this curiosity of neuroanatomy that first, the brain is split in an almost like for like mirror image, but with some significant differences based on functional specializations, i.e., McGilchrist (2019) likens the left side as concerned with "what", and the right side with "how", and second, the ubiquity of this architecture across the animal kingdom.

Having two functioning "brains" in a head allows animals, which always need to be in a state of alertness to danger, to have one side sleep while the other half is awake, and then swap over (e.g., in dolphins). However, given the ubiquity of lateralization in the animal kingdom, while the "one-half-asleep, other-half-awake" feature is not ubiquitous suggests this feature evolved after the presence of lateralization had evolved for other reasons.

The question naturally arises whether brain lateralization is a necessary feature to enable intelligence – that for intelligence to occur the brain needs to be split into hemispheres.

According to the research noted in Section 3.11 by Troy Shinbrot and Wise Young (2008) brain lateralization may have evolved to facilitate the brain's 3D visual processing of 2D sensory inputs. In evolution of the animal kingdom vision is a feature that pre-dates the first occurrence of intelligence. Intelligence may have arisen in evolution exploiting the brain structure that primarily satisfied the needs for vision. This also supports the view that visual thinking is historically the first intelligence method that evolved, that facilitated the later evolution of higher intelligent thinking – verbal thinking and fully conscious thinking.

Therefore, I speculate that brain lateralization is the starting point to create visual processing in animal brains leading to the first levels of intelligence flowing from non-self-aware consciousness.

I further speculate that the left and right hemispheres conduct an internal "dialogue" at a sub-conscious but high level. I'm not aware of any AI simulations that consider brain lateralization and suggest that this is investigated. I note that two neural networks that work together during training are deep-level generative transformers (Goodfellow et al., 2014). Having two functioning "brains" in continual communication but with different skills to offer may well be an asset in surviving in a dangerous world full of creatures that want to eat you. The weighing of evidence as seen by two perspectives of the same external stimuli can lead to superior decision making. Evolution can then select for these features, encouraging greater capacities for thinking and the first appearance of intelligence may have occurred as the result of a debating society between the two halves of the brain.

Given that the animal kingdom is populated by creatures that lack verbal thinking then the form of the suggested dialogue between brain halves is expressed by neural codes or neural ensemble oscillations (see Section 3.16) at a fundamental level and at a higher abstract level based on visual thinking.

The chatty brain, continually active whether awake or asleep, is a feature we noted in Section 3.14. Internal messaging dominates the brain while external stimuli count for less than 5% of total energy consumed by the brain. Stanislas Dehaene (2014) writes: "The nervous system primarily acts as an autonomous device that generates its own thought patterns. Even in the dark, while we rest and 'think of nothing', our brain constantly produces complex and ceaselessly changing arrays of neuronal activity".

An interesting piece of research was conducted at Meta by Mike Lewis et al. (2017) whereby two chatbots were allowed to converse with each other over tasks requiring negotiations between them and soon developed a coded "language" that they could interpret but looked incoherent to a human. It makes sense that machine to machine communication would be radically different from the kind of dialogues that humans conduct with each other. Which raises the question of what kind of "dialogue" and at what level of abstraction would the two halves of the brain conduct their communication.

10.6 MEMORY IN HLAI SYSTEMS

Memory serves to store information, and we have seen the complexity of human memory in Section 3.10. To our best understanding the human brain recreates memories from stored fragments, whereas computers can store exact memories and retrieve them accurately. The question is whether the molecular mechanics of human memory is in any manner necessary for intelligence. My guess is that we can short-circuit trying to build an exact replica of human memory at a molecular level and replace it with an engineered computer-based solution. This will require unraveling the mystery of how memorized abstract ideas and thoughts relate to neural activity. For all the advances of research in understanding memory by Eric Kandel (2006) and others, the gap between the granular activity and higher abstractions remains to be solved.

We should also be replicating the plasticity of local memory at the synapses, i.e., the ability to learn new information without forgetting learned knowledge, and to manage the transition between short term and long-term memory, which prevents memory from being overloaded with trivial data.

Hyperdimensional computing (HDC, see Section 8.8) offers advanced forms of semantic storage and retrieval with almost limitless capacity, while also enabling information and concepts to be combined and manipulated. HDC could form a foundation for an HLAI memory system. Memory not only allows recollection of the past to inform present action, but also allows the future to be imagined, and support the visualization of how things can be different is an essential skill of intelligence.

Combining electronic memory and storage with a human-like HLAI concepts makes an ideal hybrid system: $(H)^2LAI$.

10.7 THE EXECUTIVE SEAT IN THE BRAIN VERSUS DIFFUSE DECISION MAKING

There are many autonomous systems in the brain that don't require our conscious decision making. A view today sees the brain as a continuous prediction machine that only lights up our conscious decision making when a prediction turns out wrong due to unexpected events and circumstances. In society our organizations are typically structured as top-down hierarchies, with the CEO making the big decisions. There are alternative management structures, though uncommon, such as holocracy, a self-organizing system – see Brian Robertson (2015). The name holocracy was derived by Robertson from Arthur Koestler's concept of holons (Greek for whole) units that are autonomous and together form a "holarchy", but holocracy, a business management system, is formed by volunteers creating peer group circles devoting their expertise to a particular task or function, and the circle members vote others in or out. Some circles have more power than others depending on the criticality of their function. The concept has some overlap with how highly distributed large-scale open source software projects are organized, the editorship of Wikipedia content also has some parallels.

Where are decisions made in the brain? Is it a central location or diffuse? How would it work? The body has an autonomous system to deal with immediate life critical and safety critical decisions, such as reflex reactions, and habituated skills are performed unconsciously, so the concern here is with decisions that rise into our consciousness.

At one time Francis Cricks and Christof Koch suggested the claustrum as the executive seat in the brain, but according to research by Maxwell Madden et al. (2022) the claustrum acts more as a major telephone junction than a point of executive control based on animal experiments that disable the center. Through fMRI Redinbaugh et al. (2019) identify conscious activity as occurring through interactions involving central lateral thalamus and fronto-parietal cortex. Another candidate mentioned by Temple Grandin (2022) is the periaqueductal gray.

An alternative model, where decision making is more diffuse, could resemble holocracy rather than top-down management – see Figure 10.3. To implement diffuse decision-making ensembles of neurons would light up if triggered by patterns of stimuli, whether responding to external sensory data or internal imagination acting as stimuli and would compete with

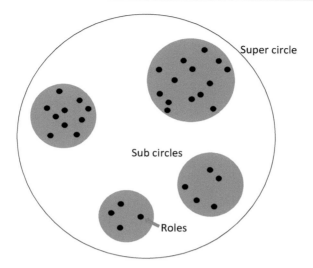

Figure 10.3 Holocracy: diffuse decision making.

Source: E. M. Azoff, based on Robertson (2015).

other ensembles. The quickest and strongest ensemble would be in a prime position to receive neuromodulators that reinforce their signals which in turn would inhibit other ensembles – effectively making that strongest ensemble the decision maker in the typical 100–200 ms window, and the cycle continues with the next thought or sensory input. The idea is to see whether neural ensembles could form holocracy circles, self-organizing to perform expert decision making.

Chapter 11

Methods to build HLAI

11.1 INTRODUCTION

Neuroscience research findings are growing at an extraordinary rate, increasing our volume of data points. Pulling these together into theories and abstractions is also extraordinarily difficult. While neuroscientists struggle to make sense of all the data, AI researchers are stuck in what appear to be primitive constructs when compared with the complexity of the human or even lesser animal brain.

This book suggests three possible avenues forward:

1 Evolutionary algorithms to grow learning systems.
2 Simulation of an ALAI system.
3 Simulation of an HLAI system.

The trick is to learn how the brain thinks and learns without needing to do what the brain does to achieve that, i.e., manage a life support system.

The recommendation here is to simulate an A/HLAI model. Simulation of a brain model can be left to run as long as necessary, forever if it has unlimited power supply. A simulation is not an inferior construct, it can create a virtual world that is end unto itself. From an AI perspective, the computer game Creatures, an artificial life simulation by Steve Grand, is an excellent example of the concept.

If we have understood all the mechanisms in the brain and created a computer version inside a simulation program, then when switched on it should emulate the living brain. There is much to decode in the human brain. David Beniaguev et al. (2021) built a DLNN to replicate the behavior of a single human brain neuron, a cortical pyramidal cell (L5PC) receiving AMPA and GABAA synapses. The I/O relationship of this physical neuron was fitted onto the artificial neural network: the artificial network was trained on identical synaptic input and axonal output of the physical neuron to replicate the I/O behavior of the physical neuron. The best artificial model, the simplest DLNN that captures the I/O properties of the single

DOI: 10.1201/9781003507864-15

real neuron needed seven layers, comprising 128 neurons per layer. For an exact replication of the human brain this needs to be multiplied over 80 billion times and with neurons of different types and dendrite complexity, and different synapse/gate mechanisms.

The artificial neuron in a DLNN model is essentially a gate. The brain's neuron is a living cell that is fixed in place and fed by assistant cells and blood flow. It survives by connecting with its neighbors, a neuron that has poor connections will wither or die altogether. Processing information in a way that is relevant to the brain is the purpose of a neuron. The neuron is also a eukaryotic cell: at its core is the nucleus, a DNA-based protein factory that receives messages and produces the required proteins. These proteins are essentially packaged algorithms. This level of flexibility and versatility has not been modeled yet in AI.

On the premise that we should solve the easier challenge first, I suggest focusing on a lesser animal brain than the human brain, which I've designated ALAI and lacking self-consciousness, possessing only in-the-flow consciousness. As nature tends to repeat itself in different animals, cracking the neural code in lesser animals will help understand the neural code in the human brain. There are plenty of animal candidates with smaller number of neurons to study than the human brain: the raven with 1.2 billion and the mouse at 70 million are still highly complex, but the bee at around 1 million shows intelligent behavior (Chittka, 2022). The most studied animal in science is probably the fruit fly, Drosophila melanogaster, it has 135,000 neurons; however, its tiny brain makes experiments challenging. Eric Kandel (2006) makes the point that selecting an anatomically simple system is crucial to the success of an experiment and that invertebrate animals are a rich source of simple systems. Kandel worked with sea snails for his experiments on memory. Alan Hodgkin & Andrew Huxley were the first to fully understand the physics of the action potential in experiments conducted in the 1930s–1940s, and they chose to work with the axon of a squid, which has a relatively huge diameter of 1 mm.

Having a complete map of the neural connections in brains will prove useful in understanding brain functions. This is being carried out by the Allen Institute using transmission electron microscopy (Yin et al., 2020). In September 2023, the Allen Institute announced a project to image an entire hemisphere of a mouse brain at 120 nm resolution. The project will assess whether to proceed to map the entire mouse brain. Earlier the Allen Institute had mapped the Drosophila connectome (Deweerdt, 2019).

Neuromodulation plays an important role in the brain, for example, dopamine acts as a reward mechanism, and adrenaline triggers a fight-or-flight response. However, only a few researchers have addressed how it may affect AI models. Kenji Doya (2002) proposed a "unified theory on the roles of neuromodulators in mediating the global learning signal and meta-parameters of distributed learning mechanisms of the brain". The same neuromodulator could have different effects on neurons and their circuits, indicating the complexity of the system. Doya (2008) returned to the topic in a review paper, highlighting how dopamine and other neuromodulators could play a role in reinforcement learning theory.

Jeffrey Krichmar (2008) suggests that the effect of neuromodulation is to cause an organism to flip between exploratory mode, when the environment appears benign, and decisive mode, when threats are perceived and action needs to be taken. Research by Jei Mei et al. (2022) integrates neuromodulatory principles into deep neural networks as a first step to exploring their potential in AI. Here, I suggest building a reward mechanism around a dopamine-like mechanism, the accretion of a virtual currency that the A/ HLAI system aims to maximize when the opportunity arises.

Another phenomenon that has been observed by neuroscientists and generally neglected by AI researchers is brain wave oscillations (see Section 3.16), but in this case, neuroscientists have tended to dismiss it as an epiphenomenon. Epiphenomena are the curse of neuroscience. With so much complexity and detail occurring in a multitude of electrical and chemical processes in the brain, determining cause and effect is a major challenge when attempting to elucidate the mechanics of thought. An epiphenomenon is an effect that occurs alongside an event or phenomenon of interest but has no functional role. Observing a phenomenon of interest and deciding whether it has a functional role or is an epiphenomenon can be a major challenge. In the research work of Wolf Singer and team (see Section 8.4), it was found that by modeling ensembles of neurons rather than individual neurons in a neural network (where nodes represent ensembles), adding oscillatory characteristics improved the learning capability of the models. This gives oscillations a direct role in improving the learning behavior of neural networks, and therefore, one concludes oscillations are not an epiphenomenon but a prime phenomenon of interest.

One feature of the brain that stares at us in every neuroanatomy textbook is the left-right split of the brain into hemispheres, discussed in Sections 3.11 and 10.5. Research into brain lateralization impairments finds peculiar behavior depending on which bits in which half are missing or inactive. In Chapter 10, I speculate the following chain of links and repeat it here: brain lateralization may have evolved to take in two-dimensional visual sensory input data and translate it into a three-dimensional view of the external world. I further speculate that visual processing leads to visual thinking and in-the-flow consciousness that lacks self-awareness and this finally leads to intelligence.

11.2 THE INTELLIGENT ROBOT AS SCIENTIST

Pulling together the findings from the survey sections of this book, the essential ingredients of an intelligent robot that will lead to HLAI are suggested

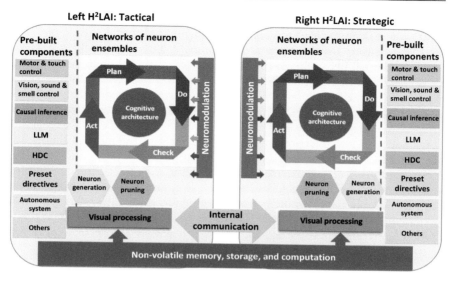

Figure 11.1 High-level conceptual model of hybrid HLAI (H²LAI).

Source: E. M. Azoff. LLM=large language model, HDC=hyperdimensional computing.

here. As stated earlier, this is running before we can walk, and I believe it would be more fruitful to focus on a lesser animal brain than a human to start with. However, in this section, I pull together key ideas which can be implemented in an ALAI system as well.

The key features of this design are as follows (see Figure 11.1):

- *1*, it is not a blueprint but a high-level conceptual list of the parts I believe are necessary for an HLAI.
- *2*, it is an H²LAI system, taking the best of engineering and AI in a hybrid design.
- *3*, while the "brain" of the system is a simulation, the robotic elements can be virtual as well, existing as a simulation, or instantiated in a physical robot moving in the real world.
- *4*, the model is a combination of pre-built components and neural network-based learning systems.
- *5*, it has a left and right brain in imitation of most living animals with any intelligence. Is this design necessary for intelligence? It may be necessary for visual processing. There may be benefit of an internal dialogue between two "minds" in decision making. For example, one half is devoted to tactical considerations, with preset directives based on such a scope, while the other half has more long-term, strategic goals, again based on preset directives acting as motivations.

At the heart of the system is an internal model that is continuously operational: the PDCA cycle. This represents an internal world (inside-out system) that can act on the external environment (virtual or real) to achieve tasks or to learn but is driven internally, rather than outside-in systems (see Section 4.2 and Buzsáki, 2019) simply reacting to some external stimuli and then halting – the limitation of most current AI systems, lacking any internal world. Constructing the PDCA system may be a task for cognitive architecture (see Section 8.2).

We will return to questions of the HLAI's purpose and motivation below, but the PDCA construction is an instantiation of the scientific learning cycle. This has been popularized in business and manufacturing circles variously as the Shewhart cycle, the Deming cycle, and in lean thinking. At its root, it is simply the scientific method stripped to essentials:

- *Plan*: Decide on some action, whether to achieve a task, or to perform a learning experiment. This requires designing the experiment.
- *Do*: Carrying out the experiment or task.
- *Check*: Monitor, observe, and study the results, halt the experiment if necessary.
- *Act*: Evaluate the results of the experiment, act on the findings, whether it is increasing knowledge, or requiring further action, or halting further action for now.

The PDCA cycle is the beating rhythm of the H²LAI system, and represents internal thought experiments, not necessarily experiments or actions to be carried out in the external environment, internal imagination is also governed by the cycle. One example of a virtual experiment is testing: the system may suggest an experiment and a virtual sub-cycle tests the idea for gaps or errors. The PDCA cycle makes the intelligent robot a proto-scientist, and this matters because the scientific method is the only method that discovers or creates knowledge. The PDCA cycle may be instantiated using cognitive architecture (see Section 8.2).

The pre-built components gather our best technology to kick-start the H²LAI system with features it should possess:

- *Motor, touch, vision, sound, and smell* (or a subset, depending on the requirements for the system): Capabilities that provide senses and facilities for moving and acting in the real world. This part of the robot includes a reflex system for protecting itself (see pre-built directives and autonomous system below).
- *Causal inference* (see Section 8.7.5): The capability to infer cause and effect where possible and have a Bayesian theory-based sense of the probabilities of event sequences which can be updated as new information is gathered.

- *Large language model* (see Section 8.6): Currently our best technology for man-machine communication.
- *Hyperdimensional computing* (see Section 8.8): The capability to store information in a semantic space and to continually update it and interact with the information it contains.
- *Preset directives*: Since the H²LAI system is not alive it has no notion of pain, designed to protect the body from harm, so a reflex system must be programmed to avoid harm to itself as well as avoid harm to others (animate or inanimate). It must be compliant with laws to avoid criminal behavior. Preset directives can also be set for the purpose and motivation of the H²LAI system: it may be an industrial system designed for specific tasks in a production environment, or it may be a research system to learn about machine intelligence. Preset directives may be re-programmed to change the behavior of the H²LAI system.
- *Autonomous system*: These are the mechanisms necessary to protect the H²LAI system from harm, including governing a reflex motor system. It also drives a clock and governs the PDCA cycle.
- *Others*: There's a host of skills and capabilities not mentioned so far that could be added to the repertoire of the H²LAI system, such as game playing, creative arts, autonomous driving, and more.
- *Neuromodulation* (see Section 3.9): Neuromodulation acts at a slower pace to neuron activation, it can provide a boost to some neurons while inhibiting others. The human brain has seven principal neuromodulators, and it is a question for further research as to how to exploit this capability in an H²LAI system. The use of dopamine in the human brain can be reflected in an H²LAI system through the reward mechanism of an internal currency, with a preset directive to maximize "earning". This internal currency can be used to drive reinforcement learning algorithms. Neuromodulation is a chemical form of communication in the brain; once it is fully understood, it is possible that in an HLAI system an engineered version can provide a design improvement using electricity.
- *Neuron generation and pruning*: As the H²LAI system is continuously learning, there needs to be a mechanism for adding neurons to neural networks, as well pruning neurons that are not performing any function.
- *Visual processing*: The brain devotes a large amount of space and energy to visual intelligence (see Section 4.6). Section 10.4 proposes that the combination of visual intelligence and the PDCA cycle gives rise to visual thinking, and this gives rise to in-the-flow consciousness that lacks self-awareness in ALAI.
- *Networks of neuron ensembles*: Neurons self-organize their responses into networks of ensembles and oscillations may play a role (see Sections 3.16 and 8.4).

- *Non-volatile memory, storage, and computation*: Electronic information technology can assist an H²LAI system, integrating it with the best of modern computing. Alex Graves et al. (2016) have shown how neural networks can read from and write to an external memory matrix, similar to a random-access memory (RAM) device.
- *Internal communication*: One possible division of labor between the left and right halves is to let the left side focus on immediate tasks while the right side can focus on longer term aims (these aims may be preset directives). Internal communication helps decision making, so if the left side suggests three possible approaches to fulfilling a task the right side may help to reduce it to one that is closest to fulfilling longer-term objectives. Visual thinking may be the method of communication.

The H²LAI system can be provided with a host of learning algorithms. I have suggested evolutionary computing learning in Figure 11.1. The system may improve itself in game playing during idle times with two party games using left versus right brain.

11.3 EVOLVING INTELLIGENT SYSTEMS

Evolutionary algorithms have the potential to be built to create ML algorithms. The resultant ML solutions could be completely novel to anything so far created. A team at Google Brain research (Real et al., 2020) has built a form of automated ML development (AutoML) that is completely free of human involvement in the architecture/design, they call this AutoML-Zero. So far, the approach has created simple ML algorithms that already exist, and the researchers aim to ramp up the project to see how far it can go to achieve novel discovery.

Given that intelligence was created in animals through Darwinian natural selection, the question arises whether it is possible to re-create that natural process in a virtual world and see how intelligence evolves by applying natural section (see Section 8.7.3), with a fast-forward acceleration in a simulation. A virtual 3D world simulated with creatures that can grow and evolve intelligence may provide clues about how real-world intelligence evolved.

Given the complexity of such a simulation, the virtual world should be built as simply as possible, with candidate creatures such as slime mold and other primitive creatures that display some intelligence. Slime mold is interesting because it shows intelligent behavior as it moves toward food sources and avoids obstacles, and as a brain-like creature devoid of a body, could make it an ideal experimental candidate (Jabr, 2012). Another simple animal to consider is the box jellyfish. This animal has no central brain but does have clusters of neurons that are light sensitive and act as primitive eyes, called rhopalia which lie on the bell part of the jellyfish. Jan Bielecki et al. (2023) found that the rhopalia nervous system acts as a learning center, performing for example, associative learning.

Even the simplest cell possesses great complexity. Moger-Reischer et al. (2023) evolved a minimal cell in an experiment of synthetic biology, where non-essential sequences are removed from an organism's genome, and compared with the original cell it was derived from as control. This engineered minimal cell had only the smallest set of genes required for cellular life. The engineered cell was allowed to evolve and after 2,000 generations regained the same fitness as the control. This research demonstrates how synthetic biology and engineering can be informed by principles of evolutionary biology and population genetics. Taking this type of synthetic/engineered biology a step further can perhaps provide a useful approach to evolving intelligent systems in simple life forms, as a real-world alternative to virtual simulations, where experiments allow the evolution of intelligence to be witnessed (there are some parallels here with Dishbrain, see Section 9.4).

11.4 LLM INTELLIGENCE AND POTENTIAL FOR HLAI

In Section 8.6.2, emergent properties in LLM models were discussed, revealing some human-like capabilities that were not specified in the training but emerged when the models were queried. If a human is tasked to stack random objects in a stable tower, we clearly invoke our understanding of gravity to achieve the task. An LLM has not been specifically trained to learn about gravity, and yet the information it has gathered in training allows it to infer properties of objects that will or will not stack. There is something in the nature of language and the vast collection of examples on the internet that allows it to infer a correct solution. Perhaps we can conclude there is some kind of "understanding" inherent purely in language, as Hinto said recently (Hinton, 2024), he believes LLM models do understand, but there are multiple levels of human understanding and LLMs are currently limited in the understanding they demonstrate.

In one research cited in Section 8.6.2, authors Bubeck et al. (2013) conclude:

> We contend that (this early version of) GPT-4 is part of a new cohort of LLMs (along with ChatGPT and Google's PaLM for example) that exhibit more general intelligence than previous AI models. ... Given the breadth and depth of GPT-4's capabilities, we believe that it could reasonably be viewed as an early (yet still incomplete) version of an artificial general intelligence (AGI) system.

In the HLAI labeling scheme introduced here, LLMs fall into the engHLAI category. Highly engineered artificial systems are only remotely inspired by the human brain; nevertheless, they achieve the highest performance of any AI system to date, and the huge sums being invested in this research are likely to yield more results.

While Part Three of this book has focused more on vision than language, language capabilities are clearly important for an AI system and hybrid systems that combine vision, language and other senses will be needed to create an HLAI system.

11.5 THE KEY ATTRIBUTES AND TESTS OF AN A/HLAI SYSTEM

In what follows, I review some of the findings in the neuroscience survey in Part One and the theory survey in Part Two and contrast these with how DLNNs represent neurons.

In Sections 3.3 and 3.4, we learned of research that showed the usual model of the action potential as originating near the soma and propagating to the synapses is only part of the story. There is also a backpropagation of the action potential that goes back to the dendrites. Moreover, action potentials can be generated at the dendrites. This rich electrical signaling is not reflected in the simple DLNN models.

It was revealed that there are three types of junctions between neurons: chemical, electrical, and field (Sections 3.8 and 3.9). Moreover, junctions can have excitatory or inhibitory effects on their neighbors and junction transmission can be one-way or bi-directional. These features add richness to the synapse that is beyond the simplified neuron junction in current DLNNs. Moreover, synapse connection strengths are highly dynamic in contrast with being fixed in DLNN after training is completed.

If there is only one neurotransmitter that we should model, it must be dopamine (Section 3.9): dopamine is the brain's reward neurotransmitter, its release creates a feeling of pleasure and reinforces the brain to repeat action that leads to its release, making it ideal for use in AI reinforcement-type learning models.

In Section 3.10, we learned that the growth and maintenance of new synaptic terminals is linked to memory persistence, i.e., long-term memory results from growth of new synapses. HLAI models must be able to grow new connections when working live in production. In DLNNs, the network is fixed after training.

The brain is a hive of activity at all times, in the wake state and asleep (Section 3.14). This does not mean all the neurons are firing, on the contrary, most neurons are inactive, and firing occurs sparsely, but with over 80 billion neurons in the human brain, a small percent active (10% according to Shoham et al. [2006]) at any given moment is still a lot of activity. It reflects an active internal world, and "interruption" from external stimuli is a small part of this activity. In contrast, DLNNs are brain dead – there is no life in these models once some input data has passed through the network. In Section 4.2, we noted Stanislas Dehaene's observation that internal brain activity overwhelms external EEG signals. Dehaene estimates that stimulus-evoked activity accounts for probably less than 5% of the energy consumed by the brain.

In Chapter 4, we encounter a key concept embraced in this book: Gyorgy Buzsáki's inside-out model. The brain's internal state creates context which impacts how the brain processes stimuli. This explains why identical stimuli can have different outcomes – the context is different. To this internal model, everything (the world, internal and external) is a mass of spiking signals and information is produced from these signals, information is not extracted from the signals (see also Chapter 6).

In Section 4.4 on animal consciousness, it was recognized that there needs to be a graded scale of animal consciousness, and Birch et al. (2020) suggested (with my two additions at the end).

- Perception – visual cognition.
- Perception – touch-based cognition.
- Evaluative richness – the spectrum of emotions.
- Unified view of self and environment integrating all senses at a point in time.
- Sense of time and time direction.
- Sound, smell, and taste-based intelligence.
- Sense of self-awareness at the highest conscious level.

In Part Two, Chapter 6, the society of mind model of independent autonomous agents (read as ensemble of neurons) working in parallel, make for diffused decision making, which we revisit as a concept in Section 10.7. The embodied AI machine (Chapter 7) is viewed by some AI researchers as an essential component of HLAI, it is through interaction with the external world that the HLAI system develops mental models.

Cognitive architectures are examined in Section 8.2, and they could play a role as pre-built components in HLAI, such as creating or facilitating an autonomous internal world.

Ensembles of neurons are also discussed in Section 3.16, where neural oscillations are found to improve learning in recurrent neural networks of neural ensembles, and this property deserves further investigation. The switch to spiking neuron models in contrast with continuous value models like DLNN is supported by neuromorphic AI researchers (see Sections 8.7.6 and 9.2) for several reasons: efficiency, low power, and rich computation dynamic possibilities.

The list below indicates what more needs to be done on the path to HLAI and the milestones yet to be achieved (there is already notable progress in several of these areas).

Near-term milestones to A/HLAI:

- Rapid learning from a few examples. Able to generalize based on seeing a few cases.
- Capability to make multiple decisions.
- Evaluates the uncertainty in a result.
- Explainable AI: revealing how a decision or result was reached.
- Achieving real-time performance: related to our progress in hardware and AI acceleration.
- Fairness and diversity, dealing with bias: more a question of eliminating these issues in the data used for training.
- Fail safe and safe closure when the AI system cannot deal with a situation – essential in safety-critical environments.
- Accumulated learning: not forgetting what has been learned.

Key attributes of an HLAI system suggested in this book are as follows:

1 An inside-out model, which means an AI brain that has an internal model of the world and applies a scientific process to discovering and learning about its environment (see Section 4.2).
2 A left-right split in the AI brain that allows it to have an internal dialogue to help decision making. Chores are divided between the halves, for example, one half can focus on immediate tasks while the other half undertakes longer-term planning.
3 An internal reward system, the equivalent of the neuromodulator dopamine, for reinforcing learning.
4 Diffuse decision making, allowing the strongest responding neuron ensembles to determine the next steps.
5 The first animal life forms grew visual apparatus and devoted considerable resources to process visual information. It makes sense to focus on building visual thinking in A/HLAI.
6 Causal reasoning: understanding cause and effect.
7 Long-term goal seeking, and the capability to autonomously set intermediate goals to achieve the long-term goal.
8 Understanding how the physical world works (schooling the AI system with science and other knowledge).
9 Ethical behavior.
10 Continuous learning, and an internal drive to accumulate knowledge and make connections.
11 Abstract thinking: the ability to abstract from details of a specific case to a general abstraction that applies more universally.

Following the suggestions in this part, the following are tests that an AI system needs to pass to be considered an HLAI solution:

1 Rapid learning from a few (one to three) examples. A typical chair with four legs and a back represents the concept of an object designed for the human body to sit on. Once that concept is understood, many objects have the capacity to be used as chairs, such as a tree log lying on the ground. Once the AI understands the idea, the notion of a chair becomes limitless – we do not want to train our AI with a million examples of chairs.

2 Some DLNN models are very deep, with hundreds of layers. The human brain has roughly six layers, but connections between neurons go between neurons in both directions between layers (reciprocal connections) and laterally within the same layer. Triangular connection patterns are also found. The degree of connection complexity in the human brain is different from typical deep learning AI models. There is also a distinction between the two action modes of neurons: excitatory and inhibitory.

3 The HLAI system must be multi-modal, able to control its mode, thus switching on learning when it needs to, applying its skills in other modes. This autonomous behavior is missing in current AI which is dependent on humans to switch on model training.

4 There is a current notion in AI that a major function of the brain is prediction: anticipate movement and change in the environment and act with this foreknowledge. Steve Grand (2003) talked of this concept and saw this as the only way a human or animal can act in good time to meet challenges or hunt, as simply reacting to events by processing sensory data in real-time and only then acting on this information is too slow, therefore the brain must be making continual predictions and acting on those predictions. In this approach, the brain is a prediction machine and acts on an internal model of the world. Predictions that relate to the physics and mechanics of the world become habituated so actions relating to them can be performed at lightning speed without thought, but when a mismatch occurs between the real world and our internal model, the mismatch jolts us into conscious awareness when we need to think more slowly and carefully, to understand where our predictions are breaking down.

5 The human brain accumulates information over time and turns it into knowledge and wisdom, and through that into creativity. The capacity of modern hyperdimensional computing systems (see Section 8.8) is virtually limitless, so we have technology to store and manipulate information, we need to integrate this capability into the rest of the AI system. We can also augment our AI system with conventional computing systems such as databases and general computation. Our AI system would have a direct capability to program conventional computers.

6 Morals, our in-built sense of what is right or wrong, and ethics, the rules of conduct we abide by in a civilized society, help navigate our

path through life. The intelligent machines we build must also possess these qualities, and with machines, it is simple to program in, or even hard-wire, rules of behavior. But of course, how to implement these rules is not always transparently clear. In ethics we know that laws can have unintended consequences, and they also change in interpretation over time: the original context may be lost when the law makers are long dead, and meaning of words change. In morality there exist ambiguities. Take the trolley problem in philosophy, see Figure 11.2 (a trolley is a US term for a UK tram or streetcar), where a well-meaning person has the dilemma of witnessing a group of children be killed by a runaway trolley or intervene by throwing a large bystander onto the track to stop the carriage and be responsible for killing an innocent person. This problem tests our values and their boundaries in the extreme and has many variants, one has the well-meaning person just press a button to divert the track and send the trolley to where only one person will die, and so does not have to commit direct murder to save the children. Teaching morals and ethics to an HLAI system and navigate its values will be a challenge.

7 Consciousness is necessary for planning and execution of actions by intelligent creatures, as discussed earlier biologists now believe that most animals have some degree of consciousness. The idea suggested in this book is that visual thinking may lead to an emergence of in-the-flow consciousness (but lacking self-awareness) and would be a first step toward simulating fully aware consciousness. ITFC is evident in some animals and would have been a precursor in human

Figure 11.2 The trolley problem studied in philosophy.

Source: E. M. Azoff.

evolution to our current consciousness capability. Lesser degrees of consciousness in varying capabilities may be characteristic of most animals.

8 Humans are driven by in-built motivations: the basics are to secure food, shelter from the elements, and procreation. Humans are also driven by curiosity and pursuit of knowledge, as well as to create works of art, literature, and music. Entertainment, games, and sports occupy our pastime. Are these essential dimensions of intelligence, or are they epiphenomena?

Chapter 12

Beyond HLAI

In Section 10.3, the ladder of increasing intelligence from ALAI to HLAI to transcendent AI was introduced, where the transcendent level has intelligent machines build ever greater levels of intelligence beyond human brain capabilities. First of all, we add to the ladder as the first rung the position of narrow AI, which is the current state of AI progress. Examples of narrow AI are all models based on DLNN, such as generative AI and LLM.

To extend the ladder, it may be useful to consider an intermediate state, between HLAI and transcendent AI: the machine-level AI (MLAI). HLAI is based on the senses we have as humans. Once achieved we could build on HLAI and augment it with new sensors that input a fuller picture of our environment. Where ALAI and HLAI have three varieties: pure engineering, animal/human-like, and hybrid – a mix of the two, MLAI is by definition an engineered variety of HLAI with sensory and limb augmentation or re-design. Figure 12.1 updates our earlier ladder of increasing intelligence accordingly.

MLAI systems may be augmented with camera "eye" sensors that are sensitive to the full spectrum of electromagnetic waves, enhancing the view of the world. AI machines with non-human like limbs or attachments that can plug into other machines and become fully connected with, say, a mode of transport. Autonomous transport, currently struggling with today's level of AI, would be fully realized with HLAI. Our buildings could be imbued with intelligence, taking today's concept of smart homes and smart cities to new levels.

I think the most transformational aspect of HLAI and beyond into transcendence is the prospect of these machines exploring galaxies. Human beings are not designed to live outside our planet and its gravitational field and protective layer of atmosphere. Astronauts who spend long periods of time in the international space station find their body changes under zero gravity; muscles are reduced, and they become weaker, more fragile. And they need protective shields and suits against harmful radiation such as cosmic rays. None of this applies to AI machines – adding radiation shields to an intelligent robot would be a natural extension.

DOI: 10.1201/9781003507864-16

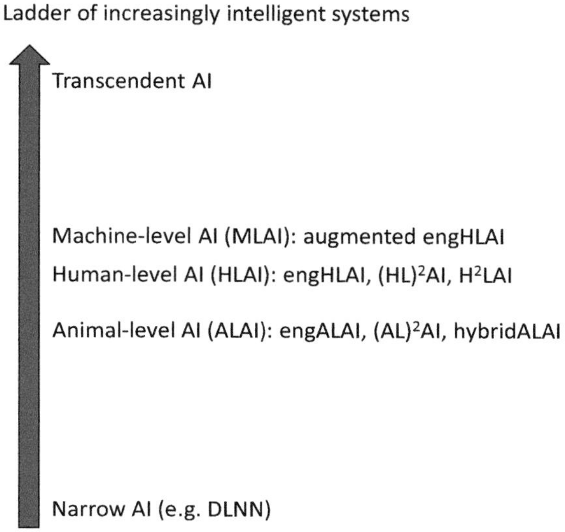

Ladder of increasingly intelligent systems

Transcendent AI

Machine-level AI (MLAI): augmented engHLAI

Human-level AI (HLAI): engHLAI, $(HL)^2AI$, H^2LAI

Animal-level AI (ALAI): engALAI, $(AL)^2AI$, hybridALAI

Narrow AI (e.g. DLNN)

Figure 12.1 Levels of increasingly intelligent AI systems – updating Figure 10.1.

Source: E. M. Azoff. $(HL)^2$=human-like, human-level; $(H)^2L$=hybrid, human-level. engHL=engineered human-level. DLNN=deep learning neural network.

I envisage transcendent AI will inherit the universe. Unlike humans, it could live forever, refreshing its mechanical and electronic body by uploading its brain into a new clone. It could hibernate while traveling in a spaceship to distant galaxies and discover new planets and seek out intelligent life forms. Given the possibility of uploading the transcendent AI into a clone, the applications expand: an AI machine could replicate itself in multiple copies. The uploading process could be modulated onto an electromagnetic wave, which traveling at the speed of light could be sent to a distant planet and create a clone separated by light years (the cloning apparatus would need to be ready set up at the receiving end). Time will not have the same meaning to transcendent AI machines as it does to humans. Our limited lifespan on Earth gives humans a sense of urgency in our activities, from which transcendent AI will be free.

Epilogue

In the middle of writing this book, on November 30, 2022, OpenAI made available to the public ChatGPT (based on LLM GPT-3.5). It was a breakthrough in public embracing of AI technology, despite occasional "hallucinations" in which ChatGPT inserted fictional responses into otherwise intelligible answers. From an HLAI perspective, what is interesting about generative AI and LLMs is their emergent properties, being able to perform various reasoning tasks, albeit with relatively poor scores (but they are getting better) but astonishing that such a system could show any promise in reasoning at all! Afterall, these systems are trained to produce coherent language, not specifically to perform reasoning. What is going on inside this artificial brain? At the time of writing, no one knows for sure as an LLM is a black box. Geoffrey Hinton conjectures that deep layers in the model can extract high-level concepts from language that lead to these emergent capabilities and says LLMs have "understanding".

For some years, I have drawn Figure 1.2 with the current "State-of-the-art in AI" line at the "Narrow AI" mark, but after ChatGPT I moved the current line a little to the right, toward HLAI. It is progress, and I expect further research will yield consistent and reliable reasoning capabilities in these models that will be an important milestone in AI. In my view to go further toward HLAI will require adoption of at least some of the concepts discussed in this book, based on neuroscience.

Appendix

Notes on the scientific method

Scientists are not taught "the scientific method" – they learn how to do scientific research through their education and subsequent practice. The people most concerned with scientific methodology tend to be philosophers of science, and there appears a consensus among them that there is no *the* scientific method, it's a matter of continuing debate. What follows is a brief overview of the topic.

Lucio Russo (2000) in *The Forgotten Revolution* points out how Archimedes and Euclid around 300 BC first gave rise to what we would recognize as science and its method, before it was reborn in Europe. Key historical figures in establishing the modern scientific method are Francis Bacon (1561–1626), who formulated principles of inductive reasoning based on observation of events and experiments, and Galileo Galilei (1564–1642), whose skills as a scientist revealed nature's hidden rules.

In recent times, Karl Popper's concept of falsifiability made clear that a theory that encompassed everything and always had an explanation for every discovery *after* it was discovered, is pretty useless as it has zero predictive powers, whereas a scientific theory needs to be in principle falsifiable by a reproducible experiment or observation. Thomas Kuhn and Paul Feyerabend talked about how science is a human endeavor (with all the foibles of humans) and gave an alternative view as to how it progresses in practice, and rather differently from Popper's rationalization; there is a lot more creativity and imagination in coming up with theories in the first place, and these can also become entrenched as dogma in the hierarchy of academia, ready for the next generation to topple over as evidence of their weaknesses accumulates: this may result in an example of Kuhn's paradigm shift.

Feyerabend had picked up a notion of Popper's who said in his lectures at LSE (Preston, 1997) that "I am a Professor of Scientific Method – but I have a problem: there is no scientific method. However, there are some simple rules of thumb, and they are quite helpful". This eventually led to Feyerabend's book *Against Method* (for which he was vilified, in some aspects Feyerabend subsequently agreed).

For a modern exposition of the scientific method, see Hugh Gauch (2012). Gauch talks of science's methodology as comprising general principles of scientific method with specialized techniques in each practicing discipline. Gauch notes his view concurs with position papers published by the American Association for the Advancement of Science and other science institutions.

For the purpose of defining a scientific process that can be embedded in an HLAI system, I provide my take on Popper's "simple rules of thumb" and Gauch's general principles. This is a process of learning to learn. So, the scientific method is a process to discover or create knowledge based on observation and experiments, whether virtual (on computers or in the mind) or physical, based on logical reasoning, often supported by mathematics, encompassing deductive, inductive, and abductive inference. It is a continuous process as follows, where failure at any step means going back to previous steps:

1 Start with an observation, a question, or a proposition.
2 Perform research (observations and/or experiments).
3 Formation of a hypothesis with explanatory, predictive, and falsifiable attributes.
4 Further tests and research.
5 Analysis of accumulated data.
6 Verification, validation, and conversion of hypothesis into a theory. Conclusions are backed by evidence or noted as conjectures.
7 Repeat cycle.

In addition, there are guiding principles such Occam's razor (principle of parsimony: choose the simpler of two competing theories that fit the facts) and where mathematics comes in (it is the language of science) scientists typically prefer selecting between competing theories that fit the facts on the basis of their aesthetic beauty/simplicity. On the latter principle, Sabine Hossenfelder in *Lost in Math* (2020, subtitled: *How beauty leads physics astray*) questions whether beauty is a valid criterion in modern physics theories, particularly theories that are currently beyond experimental verification.

Finally, to note the importance *to* the scientific method, the evolution of monitoring and measuring instruments and devices that allow science and engineering to progress. The increasing precision of measurements has directly led to progress in science and engineering. In neuroscience, there has been a recent revolution in non-destructive instrumenting of the brain, including electrophysiological sensing and scanning technology such as electroencephalogram (EEG), functional magnetic resonance imaging (fMRI), positron emission tomography (PET), latest generation magnetoencephalography (MEG), event-related potential (ERP), and optogenetics which can control neurons with light. There has also been progress in the recording probes used in the human cortex, for example, by Neuropixels (Paulk et al., 2022).

Glossary

ALAI	animal-level AI
A/HLAI	animal- or human-level AI
CPU	central processing unit
DLNN	deep learning neural network
EEG	electroencephalogram
engHLAI	engineered HLAI
ERP	event-related potential
fMRI	functional magnetic resonance imaging
GPGPU	general programming on graphics processing unit
GPU	graphics processing unit
GWT	global workspace theory
HDC	hyperdimensional computing
HLAI	human-level AI
(HL)^{2}AI	human-like human-level AI
H^{2}LAI	hybrid human-level AI
HTM	hierarchical temporal memory
IIT	integrated information theory
ITFC	in-the-flow consciousness with lack of self-awareness
LLM	large language model
MEG	magnetoencephalography
MLAI	machine-level AI
NCC	neural correlates of consciousness
PET	positron emission tomography

References

Ahmad, S., and Hawkins, J. (2015). Properties of sparse distributed representations and their application to Hierarchical Temporal Memory. Online arXiv:1503.07469.

Ahmad, S., and Hawkins, J. (2016). How do neurons operate on sparse distributed representations? Online arXiv:1601.00720.

Amari, S. (1967). A theory of adaptive pattern classifiers. *IEEE Transaction on Electronic Computers*, *EC*, 16: 299–307.

Amit, E. et al. (2017). An asymmetrical relationship between verbal and visual thinking. *Neuroimage*, 152: 619–627.

Amit, S., A., Roy, K., and Gaur, M. (2023). Neurosymbolic AI - Why, What, and How. Online arXiv:2305.00813v1 [cs.AI].

Anastassiou, C. A., and Koch, C. (2015). Ephaptic coupling to endogenous electric field activity. *Current Opinion in Neurobiology*, 31: 95–103.

Anderson, J. A., and Rosenfeld, E., eds. (1989). *Neurocomputing, Foundations of Research*. Cambridge, MA: MIT Press.

Anderson, J. A., and Rosenfeld, E., eds. (2000). *Talking Nets*. Cambridge, MA: MIT Press.

Anderson, J. R. (2007). *How Can the Human Mind Occur in the Physical Universe?* New York, NY: Oxford University Press.

Andrade-Talavera, Y., Fisahn, A., and Rodríguez-Moreno, A. (2023). Timing to be precise? An overview of spike timing-dependent plasticity, brain rhythmicity, and glial cells interplay within neuronal circuits. *Molecular Psychiatry*, 28: 2177–2188.

Araque, A. et al. (1999). Tripartite synapses. *Trends in Neuroscience*, 22: 208–215.

Arbib, M. A. (2012). *How the Brain Got Language*. New York, NY: Oxford University Press.

Arnheim, R. (1969). *Visual Thinking*. Berkley: University of California Press.

Azoff, E. M. (1994). *Neural Network Time Series Forecasting of Financial Markets*. London: John Wiley & Sons.

Baas, P. W., Rao, A. N., Matamoros, A. J., and Leo, L. (2016). Stability properties of neuronal microtubules. *Cytoskeleton (Hoboken)*, 73: 442–460.

Baars, B. J. (2021). *On Consciousness*. New York, NY: The Nautilus Press.

Baddeley, A. D. (2010). Working memory. *Current Biology*, 4: R136–R137.

Baldwin, J. M. (1896). A new factor in evolution. *The American Naturalist*, 30: 441–451. Online: www.jstor.org/stable/2453130.

Barrett, L. F. (2021). *Seven and a Half Lessons about the Brain*. London: Picador.

Barron, A. B., and Klein, C. (2016). What insects can tell us about the origins of consciousness. *PNAS*, 113: 4900–4908.

Bassett, D. S., and Bullmore, E. T. (2017). Small-world brain networks revisited. *Neuroscientist*, 23: 499–516. Online: doi:10.1177/1073858416667720.

Bellec, G. et al. (2020). A solution to the learning dilemma for recurrent networks of spiking neurons. *Nature Communications*, 11: 3625. Online: doi:10.1038/s41467-020-17236-y.

Bendor, D., and Wilson, M. A. (2012). Biasing the content of hippocampal replay during sleep. *Nature Neuroscience*, 15: 1439–1444.

Beniaguev, D. Segev, I., and London, M. (2021). Single cortical neurons as deep artificial neural networks. *Neuron*, 109: 2727–2739.e3. Online: doi:10.1016/j.neuron.2021.07.002.

Bennett, M. V. L., and Zukin, R. S. (2004). Electrical coupling and neuronal synchronization in the mammalian brain. *Neuron*, 41: 495–511.

Berkes, P., Orbán, G. Máté Lengyel, G. M., and Fiser, J. (2011). Spontaneous cortical activity reveals hallmarks of an optimal internal model of the environment. *Science*, 331: 83–87.

Bergles, D. E., Roberts, J. D. B., Somogyi, P., and Jahr, C. E. (2000). Glutamatergic synapses on oligodendrocyte precursor cells in the hippocampus. *Nature*, 405: 187–191.

Bermudez, J. L. (2023). *Cognitive Science*, 5th edn. Cambridge: Cambridge University Press.

Beyeler, M. et al. (2019). Neural correlates of sparse coding and dimensionality reduction. *PLoS Computational Biology*, 15: e1006908. doi:10.1371/journal.pcbi.1006908.

Billaudelle, S., and Ahmad, S. (2016). Porting HTM Models to the Heidelberg Neuromorphic Computing Platform. Online: arXiv:1505.02142v2 [q-bio.NC].

Bielecki, J., Dam Nielsen, S. K., Nachman, G., and Garm, A. (2023). Associative learning in the box jellyfish Tripedalia cystophora. *Current Biology*, 33: 4150–4159.e5. Online: doi:10.1016/j.cub.2023.08.056.

Birch, J. et al. (2020). Dimensions of animal consciousness. *Trends in Cognitive Sciences*, 24: 789–801.

Bloch, G., Barnes, B. M., Gerkema, M. P., and Helm, B. (2013). Animal activity around the clock with no overt circadian rhythms. *Proceedings of the Royal Society B*, 280: 20130019. Online: doi:10.1098/rspb.2013.0019.

Boldrini, M. et al. (2018). Human hippocampal neurogenesis persists throughout aging. *Cell Stem Cell*, 22: 589–599.

Bommasani, R. et al. (2022). On the Opportunities and Risks of Foundation Models. Online arXiv:2108.07258v3 [cs.LG].

Bonilla, L., Gautrais, J., Thorpe, S., and Masquelier, T. (2022). Analyzing time-to-first-spike coding schemes. *Frontier Neuroscience*, 16: 971937. Online: doi:10.3389/fnins.2022.971937.

Botvinick, M. et al. (2020). Deep reinforcement learning and its neuroscientific implications. *Neuron*, 107: 603–616. Online: doi:10.1016/j.neuron.2020.06.014.

Brandman, T., Malach, R., and Simony, E. (2021). The surprising role of the default mode network in naturalistic perception. *Communication Biology,* 4: 79. Online: doi:10.1038/s42003-020-01602-z.

Bray, D. (2009). *Wetware.* New Haven, CT: Yale University Press.

Brito da Silva, L. E., Elnabarawya, I., and Wunsch II, D. C. (2019). A Survey of Adaptive Resonance Theory Neural Network Models for Engineering Applications. Online: arXiv:1905.11437v1 [cs.NE]

Brogaard, B. (2012). Kim Peek, the Real Rain Man. *Psychology Today,* online: https://www.psychologytoday.com/gb/blog/the-superhuman-mind/201212/kim-peek-the-real-rain-man.

Brzosko, Z., Mierau, S. B., and Paulsen, O. (2019). Neuromodulation of spike-timing-dependent plasticity: Past, present, and future. *Neuron,* 103: 563–581.

Bubeck, S. et al. (2023). Sparks of Artificial General Intelligence. Online: arXiv:2303.12712v5 [cs.CL].

Buchanan, J., da Costa, N. M., and Cheadle, L. (2023). Emerging roles of oligodendrocyte precursor cells in neural circuit development and remodeling. *Trends in Neuroscience,* 5: S0166-2236(23)00132-7.

Buckner, R. L. (2013). The brain's default network: Origins and implications for the study of psychosis. *Dialogues Clinical Neuroscience,* 15: 351–358. Online: doi:10.31887/DCNS.2013.15.3/rbuckner.

Bullock, T. H., Bennett, M. V., Johnston, D., Josephson, R., Marder, E., and Fields, R. D. (2005). Neuroscience. The neuron doctrine, redux. *Science,* 310: 791–793.

Buzsáki, G., and Draguhn, A. (2004). Neuronal oscillations in cortical networks. *Science,* 304: 1926–1929.

Buzsáki, G., and Moser, E. I. (2013). Memory, navigation and theta rhythm in the hippocampal- entorhinal system. *Nature Neuroscience,* 16: 130–138.

Buzsáki, G., and Llinas, R. (2017). Space and time in the brain. *Science,* 358: 482–485.

Buzsáki, G., and Tingley, D. (2018). Space and time. *Trends in Cognitive Sciences,* 22: 853–869.

Buzsáki, G. (2011). *Rhythms of the Brain.* New York, NY: Oxford University Press.

Buzsáki, G. (2019). *The Brain from Inside Out.* New York, NY: Oxford University Press.

Cambridge Declaration on Consciousness (2012). Online: https://fcmconference.org/img/CambridgeDeclarationOnConsciousness.pdf.

Camina, E. and Güell, F. (2017). The neuroanatomical, neurophysiological and psychological basis of memory: Current models and their origins. *Frontier Pharmacology,* 8: 438. doi:10.3389/fphar.2017.00438.

Campenot, R. B. (2016). *Animal Electricity.* Cambridge, MA: Harvard University Press.

Cao, Y., and Grossberg, S. (2012). Stereopsis and 3D surface perception by spiking neurons in laminar cortical circuits. *Neural Networks,* 26: 75–98. Online: https://sites.bu.edu/steveg/files/2016/06/CaoGroTR2011.pdf.

Carpenter, G. A., and Grossberg, S. (1987). A massively parallel architecture for a self-organizing neural pattern recognition machine. *Computer Vision, Graphics, and Image Processing,* 37: 54–115. Online: https://sites.bu.edu/steveg/files/2016/06/CarGro1987CVGIP.pdf.

Carpenter, P. A., Just, M. A., and Shell, P. (1990). What one intelligence test measures: A theoretical account of the processing in the raven progressive matrices test. *Psychological Review*, 97: 404–431.

Carlo C. N., and Stevens (2013). Structural uniformity of neocortex, revisited. *PNAS*, 110: 1488–1493.

Carandini, M., and Heeger, D. J. (2012). Normalization as a canonical neural computation. *Nature Reviews Neuroscience*, 13: 51–62.

Cataldo, D. M., Migliano, A. B., and Vinicius, L. (2018). Speech, stone toolmaking and the evolution of language. *PLoS One*, 13(1): e0191071. Online: doi:10.1371/journal.pone.0191071.

Chelini, G. et al. (2018). The tetrapartite synapse. *European Psychiatry*, 50: 60–69.

Chittka, L. (2017). Bee cognition. *Current Biology*, 27: R1037–R1059.

Chittka, L. (2022). *The Mind of a Bee*. Princeton, NJ: Princeton University Press.

Cho, W. H., Barcelon, E., and Lee, S. J. (2016). Optogenetic glia manipulation. *Experimental Neurobiology*, 25: 197–204.

Christensen, D. V. et al. (2022). 2022 Roadmap on neuromorphic computing and engineering. *Neuromorphic Computing and Engineering*, 2: 022501. Online: doi:10.1088/2634-4386/ac4a83.

Clark, A. (2015). *Surfing Uncertainty*. New York, NY: Oxford University Press.

Cobbe, K. et al. (2021). Training Verifiers to Solve Math Word Problems. Online: arXiv:2110.14168v1 [cs.LG].

Connors, B. W., and Long, M. A. (2004). Electrical synapses in the mammalian brain. *Annual Review of Neuroscience*, 27: 393–418.

Corballis, M. C. (2003). From mouth to hand: Gesture, speech, and the evolution of righthandedness. *Behavioral and Brain Science*, 26: 198–208.

Corballis, M. C. (2014). Left brain, right brain: Facts and fantasies. *PLoS Biology*, 12: e1001767. doi:10.1371/journal.pbio.1001767.

Cui, Y., Ahmad, S., and Hawkins, J. (2016). Continuous online sequence learning with an unsupervised neural network model. *Neural Computation*, 28: 2474–2504.

Cui, Y., Ahmad, S., and Hawkins, J. (2017). The HTM spatial pooler-a neocortical algorithm for online sparse distributed coding. *Frontier in Computational Neuroscience*, 11: 111.

Czégel, D. et al. (2021). Novelty and imitation within the brain: A Darwinian neurodynamic approach to combinatorial problems. *Nature Science Report*, 11: 12513. Online: doi:10.1038/s41598-021-91489-5.

Davies, M. et al. (2021). Advancing neuromorphic computing with Loihi: A survey of results and outlook. *Proceeding of IEEE*, 109: 911–934.

De Bono, E. (1967). *The Use of Lateral Thinking*. London: Jonathan Cape.

Dehaene, S. (2014). *Consciousness and the Brain*. New York, NY: Penguin Books.

Dehaene, S., Nicolas, M., Cohen, L., and Wilson, A. J. (2004). Arithmetic and the brain. *Current Opinion in Neurobiology*, 14: 218–224.

Dempster, A. P., Laird, N. M., and Rubin, D. B. (1977). Maximum likelihood from incomplete data via the EM algorithm. *Journal Royal of Statics Society Series B*, 39: 1–38.

Deweerdt, S. (2019). Deep connections. *Nature*, 571: S6–S8.

Dodig-Crnkovic, G. (2021). Natural Computational Architectures for Cognitive Info-Communication. Online arXiv:2110.06339 [q-bio.NC].

Dong, A., Liu, S., and Li, Y. (2018). Gap junctions in the nervous system. *Frontier in Cellular Neuroscience*, 12: 320–328.

Doya, K. (2002). Metalearning and neuromodulation. *Neural Networks*, 15: 495–506.

Doya, K. (2008). Modulators of decision making. *Nature Neuroscience*, 11: 410–416.

Droria, I. et al. (2022). A neural network solves, explains, and generates university math problems by program synthesis and few-shot learning at human level. *PNAS*, 119: e2123433119. Online: doi:10.1073/pnas.2123433119.

Du, J. et al. (2023). Within-Individual Organization of the Human Cerebral Cortex. Online: bioRxiv preprint. doi:10.1101/2023.08.08.552437.

Duan, J. et al. (2022). A Survey of Embodied AI. Online: arXiv:2103.04918 [cs.AI].

Eagleman, D. M. et al. (2005). Time and the brain. *The Journal of Neuroscience*, 25: 10369–10371.

Eagleman, D. (2021). *Livewired*. Edinburgh: Canongate Books.

Eagleman, D. M., and Vaughn, D. A. (2021). The defensive activation theory. *Frontier Neuroscience*, 15: 632853. Online: doi:10.3389/fnins.2021.632853.

Edelman, G. M. (1987). *Neural Darwinism*. New York, NY: Basic Books.

Edelman, G. M., and Tononi, G. (2000). *Consciousness*. London: Penguin Books.

Edelman, G. M., and Gally, J. A. (2013). Reentry: A key mechanism for integration of brain function. *Frontiers in Integrative Neuroscience*, review article, 7: 63. Online: doi:10.3389/fnint.2013.00063.

Eden, J. et al. (2022). Principles of human movement augmentation and the challenges in making it a reality. *Nature Communications*, 13: 1345. Online: doi:10.1038/s41467-022-28725-7.

Effenberger, F. et al. (2023). The functional role of oscillatory dynamics in neocortical circuits: a computational perspective. Online: https://www.biorxiv.org/content/biorxiv/early/2023/09/01/2022.11.29.518360.full.pdf

Einstein, A. Correspondence with Max Wertheimer, 1959. Quoted in: Pais, A. (1982). Subtle is the Lord: The Science and the Life of Albert Einstein. Oxford: Oxford University Press.

Eliasmith, C. (2013). *How to Build a Brain*. New York, NY: Oxford University Press.

Eliasmith, C., and Furlong, P. M. (2021). Continuous then discrete: A recommendation for building robotic brains. *Blue Sky Papers, 5th Conference on Robot Learning (CoRL 2021, London), PMLR*, 164: 1758–1763.

Elston, G. N. (2003). Cortex, cognition and the cell: new insights into the pyramidal neuron and prefrontal function. *Cerebral Cortex*, 13: 1124–1138. Online: doi:10.1093/cercor/bhg093.

Emery, N. (2016). *Bird Brain*. Princeton, NJ: Princeton University Press.

Ernoult, M. et al. (2020). Equilibrium Propagation with Continual Weight Updates. Online: arXiv:2005.04168v1 [cs.NE].

Faber, D. S., and Pereda, A. E. (2018). Two forms of electrical transmission between neurons. *Frontier in Molecular Neuroscience*, 11: 427. Online: doi:10.3389/fnmol.2018.00427.

Federmeier, K., and Benjamin, A. (2005). Hemispheric asymmetries in the time course of recognition memory. *Psychonomic Bulletin & Review*, 12: 993–998.

Federmeier, K. (2007). Thinking ahead: The role and roots of prediction in language comprehension. *Psychophysiology*, 44: 491–505.

Federmeier, K. D., Wlotko, E. W., and Meyer, A. M. (2008). What's "right" in language comprehension. *Lang Linguist Compass*, 2: 1–17.

Fedor, A. et al. (2017). Cognitive architecture with evolutionary dynamics solves insight problem. *Frontier in Psychology*, 8: 427. Online: doi:10.3389/fpsyg.2017.00427.

Felleman, D. J., and Van Essen, D. C. (1991). Distributed hierarchical processing in the primate cerebral cortex. *Cereb Cortex*, 1: 1–47. Online: doi:10.1093/cercor/1.1.1-a. PMID: 1822724.

Fernando, C., Szathmáry, E., and Husbands, P. (2012). Selectionist and evolutionary approaches to brain function: A critical appraisal. *Frontiers in Computational Neuroscience*, 6: 24. Online: doi:10.3389/fncom.2012.00024.

Fernando, C. et al. (2018). Meta-Learning by the Baldwin Effect. Online arXiv:1806.07917v2 [cs.NE].

Fleming, S. M. et al. (2023). The Integrated Information Theory of Consciousness as Pseudoscience. *PsyArXiv Preprints*. Online: doi:10.31234/osf.io/zsr78.

Forrest, M. D. (2014). The sodium-potassium pump is an information processing element in brain computation. *Frontiers in Physiology*, 5: 472. Online: doi:10.3389/fphys.2014.00472.

Foster, R. (2022). *Life Time*. London: Penguin Books.

Fouts, R. H., Fouts, D. M., Van Cantfort, T. E. (1989). In B. Gardner, R. Allen, et al. (eds.), *Teaching Sign Language to Chimpanzees*. New York: SUNY Press, 281–282.

Franklin, S. (1999). *Artificial Minds*. Cambridge, MA: MIT.

Frasnelli, E. (2013). Brain and behavioral lateralization in invertebrates. *Frontiers in Psychology*, Online: doi:10.3389/fpsyg.2013.00939.

Friston, K. (2003). Learning and inference in the brain. *Neural Networks*, 16: 1325–1352.

Friston, K., Kilner, J., and Harrison, L. (2006). A free energy principle for the brain. *Journal of Physiology*, 100: 70–87.

Friston, K. (2010). The free-energy principle: A unified brain theory? *Nature Reviews Neuroscience*, 11: 127–138.

Fusi, S., Miller, E. K., and Rigotti, M. (2016). Why neurons mix: High dimensionality for higher cognition. *Current Opinion in Neurobiology*, 37: 66–74.

Garcez, A. d'A., and Lamb, C. L. (2020). Neurosymbolic AI: The 3rd Wave. Online arXiv:2012.05876 [cs.AI].

Garcia, K. E., Kroenke, C. D., and Bayly, P. V. (2018). Mechanics of cortical folding: Stress, growth and stability. *Philosophical Transactions of the Royal Society B*, 373: 20170321.

Ganguli, D. et al. (2022). Predictability and Surprise in Large Generative Models. Online arXiv:2202.07785v2 [cs.CY].

Gauch, Jr, H. G. (2012). *Scientific Method in Brief*. Cambridge: CUP.

George, D. (2008). *How the Brain Might Work: A Hierarchical and Temporal Model for Learning and Recognition*. Ph.D. thesis, Stanford University.

George, D., and Hawkins, J. (2009). *Towards a Mathematical Theory of Cortical Micro-circuits*. Online: PLoS Computational Biology, 5: 1–26.

Gierer, A. (2008). Brain, mind and limitations of a scientific theory of human consciousness. *Bioessays*, 30: 499–505. Online: doi:10.1002/bies.20743.

Gilbert, C. D., and Li, W. (2013). Top-down influences on visual processing. *Nature Reviews Neuroscience*, 14: 350–363.

Ginosar, G. et al. (2021). Locally ordered representation of 3D space in the entorhinal cortex. *Nature*, 596: 404–409.

Givon, S., Samina, M., Ben-Shahar, O., and Segev, R. (2022). From fish out of water to new insights on navigation mechanisms in animals. *Behavioural Brain Research*, 419. Online: doi:10.1016/j.bbr.2021.113711. For a video of the fish in motion: Interestingengineering.com/video/this-goldfish-controls-a-robotic-tank-with-its-movement.

Goodale, M. A., and Milner, D. A. (1992). Separate visual pathways for perception and action. *Trends in Neurosciences*, 15: 20–25.

Goodfellow, I. J. et al. (2014). Generative adversarial nets. *Proceedings of the International Conference on Neural Information Processing Systems (NIPS)*, 2014: 2672–2680.

Grand, S. (2003). *Growing Up with Lucy*. London: Weidenfeld & Nicolson.

Grandin, T. (2022). *Visual Thinking*. London: Penguin Random House UK.

Graves, A. et al. (2016). Hybrid computing using a neural network with dynamic external memory. *Nature*, 538: 471–476.

Greicius, M. D., Krasnow, B., Reiss, A. L., and Menon, V. (2003). Functional connectivity in the resting brain. *Proceeding of the National Academy Science of the United States America*, 100: 253–258.

Grossberg, S. (1980). How does a brain build a cognitive code? *Psychological Review*, 87: 1–51. Reproduced in Anderson & Rosenfeld (1989).

Grossberg, S., and Versace, M. (2008). Spikes, synchrony, and attentive learning by laminar thalamocortical circuits. Online: https://sites.bu.edu/steveg/files/2016/06/GroVer2008BR.pdf.

Grossberg, S. (2013). Adaptive resonance theory. *Neural Networks*, 37: 1–47.

Grossberg, S. (2018). Desirability, availability, credit assignment, category learning, and attention. *Brain and Neuroscience Advances*, 2: 1–50. Online: doi:10.1177/2398212818772179.

Grossberg, S. (2020). A path toward explainable AI and autonomous adaptive intelligence. *Frontiers in Neurorobotics*, 14: article 36. Online: doi:10.3389/fnbot.2020.00036.

Grossberg, S. (2021). *Conscious MIND Resonant BRAIN*. New York, NY: Oxford University Press.

Hafner, A.-S., Donlin-Asp, P. G., Leitch, B., Herzog, E., and Schuman, E. M. (2019). Local protein synthesis is a ubiquitous feature of neuronal pre- and postsynaptic compartments. *Science*, 364, No. 6441.

Hafting, T., Fyhn, M., Molden, S. et al. (2005). Microstructure of a spatial map in the entorhinal cortex. *Nature*, 436: 801–806.

Hagen, S. (2012). The mind's eye. *Rochester Review*, 74: 4.

Hall, J. C., Rosbash, M., and Young, M. (2017). Nobel Prize in Physiology or Medicine awarded jointly for their discoveries of molecular mechanisms controlling the circadian rhythm. Online: www.nobelprize.org/prizes/medicine/2017/press-release.

Han, K-S. et al. (2020). Climbing fiber synapses rapidly and transiently inhibit neighboring Purkinje cells via ephaptic coupling. *Nature Neuroscience*, 23: 1399–1409.

Hagan, M. T., Demuth, H. B., Beale, M., and De Jesus, O. (2014). *Neural Network Design*, 2nd edn. Ontario: Thompson Learning. See the book website: Hagan. okstate.edu/nnd.html.

Hawkins, J., and Blakeslee, S. (2014). *On Intelligence*. New York, NY: St. Martin's Publishing.

Hawkins, J., Ahmad, S., and Cui, Y. (2017). A theory of how columns in the neocortex enable learning the structure of the world. *Frontier in Neural Circuits*, 11: 81. Online: doi:10.3389/fncir.2017.00081.

Hawkins, J. et al. (2019). A framework for intelligence and cortical function based on grid cells in the neocortex. *Frontier in Neural Circuits*, 12: 121. Online: doi:10.3389/fncir.2018.00121.

Hawkins, J. et al. (2020). *Biological and Machine Intelligence*. Release 0.4. Accessed at https://numenta.com/resources/biological-and-machine-intelligence/.

Hawkins, J. (2021). *A Thousand Brains*. New York, NY: Basic Books.

Haykin, S. (2009). *Neural Networks and Learning Machines*, 3rd edn. London: Pearson.

Hassabis, D. et al. (2017). Neuroscience-inspired artificial intelligence. *Neuron*, 95: 245–258.

Hasselmo, M. E. (2013). *How We Remember*. Cambridge, MA: MIT Press.

Hebb, D. (1949). *The Organization of Behavior*. New York, NY: Wiley.

Herculano-Houzel, S. et al. (2008). The basic nonuniformity of the cerebral cortex. *PNAS*, 105: 12593–12598.

Herculano-Houzel, S. (2016). *The Human Advantage*. Cambridge, MA: MIT Press.

Herculano-Houzel, S. (2020). Birds do have a brain cortex—and think. *Science*, 369: 1567–1568.

Hernandez-Orallo, J. (2017). *The Measure of All Minds*. Cambridge: Cambridge University Press.

Hersche, M. et al. (2023). A Neuro-vector-symbolic Architecture for Solving Raven's Progressive Matrices. Online arXiv:2203.04571v2 [cs.LG].

Hertz, J. A., Krogh, A., and Palmer, R. G. (1991). *Introduction to the Theory of Neural Computation*. Redwood, CA: Addison Wesley.

Hickok, G. (2014). *The Myth of Mirror Neurons*. New York, NY: W. W. Norton.

Hoffman, D. D. (1998). *Visual Intelligence*. New York, NY: W. W. Norton & Company.

Hole, K. J., and Ahmad, S. (2021). A thousand brains: Toward biologically constrained AI. Online: *SN Applied Sciences*, 3: 743.

Howard, M. W. (2018). Memory as perception of the past. *Trends in Cognitive Science*, 22: 124–136.

Howarth, C., Peppiatt-Wildman, C. M., and Attwell, D. (2010). The energy use associated with neural computation in the cerebellum. *Journal of Cerebral Blood Flow & Metabolism*, 30: 403–414.

Howhy, J. (2013). *The Predictive Mind*. Oxford: Oxford University Press.

Hossenfelder, S. (2020). *Lost in Math*. London: Basic Civitas Books.

Hughes, J. et al. (2021). Embodied artificial intelligence. *IOP Conference Series: Material Science and Engineering*, 1261 012001.

Hunter, K., Spracklen, L., and Ahmad, S. (2022). Two sparsities are better than one. *Neuromorphic Computing Engineering*, 2. Online: doi:10.1088/2634-4386/ac7c8a.

Huxter, J., Burgess, N., and O'Keefe, J. (2003). Independent rate and temporal coding in hippocampal pyramidal cells. *Nature*, 425: 828–832.

Hwu, T., and Krichmar, J. L. (2020). A neural model of schemas and memory encoding. *Biological Cybernetics*, 114: 169–186. Online: doi:10.1007/s00422-019-00808-7.

Innocenti, G. M., and Price, D. J. (2005). Exuberance in the development of cortical networks. *Nature Review Neuroscience*, 6: 955–965.

Isomura, T., Kotani, K., Jimbo, Y. et al. Experimental validation of the free-energy principle with in vitro neural networks. *Nature Communication*, 14: 4547. Online: doi:10.1038/s41467-023-40141-z.

Ivakhnenko, A. G., and Lapa, V. G. (1965). *Cybernetic Predicting Devices*. Yuba City: CCM Information Corporation.

Jabr, F. (2012). How brainless slime molds redefine intelligence. *Nature Scientific American*. Online: doi:10.1038/nature.2012.11811.

Johnston, W. J., Palmer, S. E., and Freedman, D. J. (2020). Nonlinear mixed selectivity supports reliable neural computation. *PLoS Computational Biology*, 16: e1007544. Online: doi:10.1371/journal.pcbi.1007544.

Jonas, E., and Kording, K. P. (2017). Could a neuroscientist understand a microprocessor? *PLoS Computational Biology*, 13: e1005268. Online: doi:10.1371/journal.pcbi.1005268.

Jones, R. M. et al. (1999). Automated intelligent pilots for combat flight simulation. *AI Magazine*, 20: 27–41.

Kagan, B. J. (2022). In vitro neurons learn and exhibit sentience when embodied in a simulated game-world. *Neuron*, 110: 3952–3969.

Kandel, E. R. (2006). *In Search of Memory*. New York, NY: W. W. Norton & Company.

Kandel, E. R., Schwartz, J. H., Jessell, T. M., Siegelbaum, S. A., and Hudspeth, A. J., eds. (2013). *Principles of Neural Science*, 5th edn. New York, NY: McGraw-Hill.

Kandel, E. R., Dudai, Y., and Mayford, M. R. (2014). The molecular and systems biology of memory. *Cell*, 157: 163–186.

Kanerva, P. (1990). *Sparse Distributed Memory*. Cambridge, MA: MIT Press.

Kanerva, P. (2009). Hyperdimensional computing. *Cognitive Computing*, 1: 139–159. Online: doi:10.1007/s12559-009-9009-8.

Kaufman, M. T. et al. (2022). The implications of categorical and category-free mixed selectivity on representational geometries. *Current Opinion in Neurobiology*, 77: 102644. Online: doi:10.1016/j.conb.2022.102644.

Kemmerer, D. (2015). Does the motor system contribute to the perception and understanding of actions? *Language and Cognition*, 7: 450–475.

Kendall, J., Pantone, R., Manickavasagam, K., Bengio, Y., and Scellier, B. (2020). Training End-to-End Analog Neural Networks with Equilibrium Propagation. Online arXiv:2006.01981v2 [cs.NE].

Kenet, T., Bibitchkov, D., Tsodyks, M. et al. (2003). Spontaneously emerging cortical representations of visual attributes. *Nature*, 425: 954–956. Online: doi:10.1038/nature02078.

Killian, N. J., Jutras, M. J., & Buffalo, E. A. (2012). A map of visual space in the primate entorhinal cortex. *Nature*, 491: 761–764. Online: doi:10.1038/nature11587.

Kilner, J. M., and Lemon, R. N. (2013). What we know currently about mirror neurons. *Current Biology*, 23: R1057–R1062.

Kim, D. et al. (2019). Spontaneously emerging patterns in human visual cortex and their functional connectivity are linked to the patterns evoked by visual stimuli. *Journal of Neurophysiology*, 124. doi:10.1152/jn.00630.2019.

Kira, S. et al. (2023). A distributed and efficient population code of mixed selectivity neurons for flexible navigation decisions. *Nature Communications*, 14: 2121. Online: doi:10.1038/s41467-023-37804-2.

Kleyko, D., Rachkovskij, D. A., Osipov, E., and Rahimi, A. (2022a). A Survey on Hyperdimensional Computing aka Vector Symbolic Architectures, Part I: Models and Data Transformations. Online arXiv:2111.06077v1 [cs.AI].

Kleyko, D., Rachkovskij, D. A., Osipov, E., and Rahimi, A. (2022b). A Survey on Hyperdimensional Computing aka Vector Symbolic Architectures, Part II: Applications, Cognitive Models, and Challenges. Online arXiv:2112.15424v2 [cs.AI].

Koch, C. (2004). *The Quest for Consciousness: A Neurobiological Approach.* Englewood, US-CO: Roberts & Company.

Kolb, B., and Fantie, B. D. (2008). In chapter 2: Development of the Child's Brain and Behavior. In C. R. Reynolds and E. Fletcher-Janzen (eds.), *Handbook of Clinical Child Neuropsychology* (2009), Berlin: Springer Science.

Kolb, B., and Gibb, R. (2011). Brain plasticity and behaviour in the developing brain. *Journal of Canadian Academy of Child Adolescent Psychiatry*, 20: 265–276.

Kolb, B., Harker, A., and Gibb, R. (2017). Principles of plasticity in the developing brain. *Development Medicine and Child Neurology*, 59: 1218–1223.

Koppenol-Gonzalez, G. V., Bouwmeester, S., and Vermunt, J. K. (2018). Accounting for individual differences in the development of verbal and visual short term memory processes in children. *Learning and Individual Differences*, 66: 29–37.

Kotseruba, I., and Tsotsos, J. K. (2020). 40 years of cognitive architectures. *Artificial Intelligence Review*, 53: 17–94.

Koulakov, A., Shuvaev, S., Lachi, D., and Zador, A. (2022). Encoding innate ability through a genomic bottleneck. *bioRxiv* preprint. doi:10.1101/2021.03.16.435261.

Kozhevnikov, M. et al. (2013). Creativity, visualization abilities, and visual cognitive style. *British Journal of Educational Psychology*, 83: 196–209.

Krichmar, J. L. (2008). The neuromodulatory system: A framework for survival and adaptive behavior in a challenging world. *Adaptive Behavior*, 16: 385–399.

Krichmar, J. L., and Hwu, T. J. (2022). Design principles of neurorobotics. *Frontiers in Neurorobotics*, 16: online article 882518.

Krupenye, C., and Call, J. (2019). Theory of mind in animals. *WIREs Cognitive Science,* 1(25): e1503. Online: doi:10.1002/wcs.1503.

Kurzweil, R. (2005). *The Singularity Is Near: When Humans Transcend Biology.* London: Duckworth.

Laird, J., and Rosenbloom, P. (1994). The Evolution of the Soar Cognitive Architecture. Online: www.researchgate.net/publication/2719989_The_Evolution_of_the_Soar_Cognitive_Architecture/citation/download.

Laird, J. E. (2022). Introduction to the Soar Cognitive Architecture. Online: arxiv. org/ftp/arxiv/papers/2205/2205.03854.pdf.

Lake, B. M., Ullman, T. D., Tenenbaum, J. B., and Gershman, S. J. (2017). Building machines that learn and think like people. *Behavioral and Brain Sciences*, 40: e253. Online: doi:10.1017/S0140525X16001837.

Land, M. F. (2014). Do we have an internal model of the outside world? *Philosophical Transaction on Royal Society B*, 369: 20130045. Online: doi:10.1098/rstb.2013.0045.

Land, M. F. (2018). *Eyes to See*. Oxford: Oxford University Press.

Laughlin, S. B., and Sejnowski, T. J. (2003). Communication in neuronal networks. *Science*, 301: 1870–1874.

Lebedev, M. A., and Nicolelis, M. A. (2017). Brain-machine interfaces. *Physiology Review*, 97: 767–837.

Le Cun, Y. (1985). A learning scheme for asymmetric threshold network. *Cognitiva*, 85: 599–604. Paris: CESTA.

Le Cun, Y., Cortes, C., and Burges, C. J. (2010). MNIST handwritten digit database. ATT Labs online: http://yann.lecun.com/exdb/mnist.

Le Cun, Y., Bengio, Y., and Hinton, G. (2015). Deep learning. *Nature*, 521: 436–444.

Leterrier, C. (2018). The axon initial segment. *The Journal of Neuroscience*, 38: 2135–2145.

Levenson, J. M., and Sweatt, J. D. (2005). Epigenetic mechanisms in memory formation. *Nature Review Neuroscience*, 6: 108–118.

Levine, D. S. (2019). *Introduction to Neural and Cognitive Modeling*, 3rd edn. New York, NY: Routledge.

Levitan, I. B., and Kaczmarek, L. K. (2015). *The Neuron*, 4th edn. New York, NY: Oxford University Press.

Lewis, M. et al. (2017). Deal or No Deal? End-to-End Learning for Negotiation Dialogues. Online arXiv:1706.05125 [cs.AI].

Liao, Q., Leibo, J. Z., and Poggio, T. (2016). How Important Is Weight Symmetry in Backpropagation? Online arXiv:1510.05067 [cs.LG].

Lieto, A. (2021). *Cognitive Design for Artificial Minds*. Abingdon, Oxon: Routledge.

Lillicrap, T. P., Cownden, D., Tweed, D. B., and Akerman, C. J. (2014). Random feedback weights support learning in deep neural networks. Online arXiv:1411.0247v1 [q-bio.NC].

Lillicrap, T. P., Santoro, A., Marris, L., Akerman, C. J., and Hinton, G. (2020). Backpropagation and the brain. *Nature Reviews Neuroscience*, 21: 335–346. Online: doi:10.1038/s41583-020-0277-3.

Llinas, R. R. (2013). The olivo-cerebellar system: A key to understanding the functional significance of intrinsic oscillatory brain properties. *Front Neural Circuits*, 7: 96. Online: doi:10.3389/fncir.2013.00096.

Lisman, J., Cooper, K., Sehgal, M., and Silva, A. J. (2018). Memory formation depends on both synapse-specific modifications of synaptic strength and cell-specific increases in excitability. *Nature Neuroscience*, 21: 309–314.

Liu, J., and Nussinov, R. (2016). Allostery: An overview of its history, concepts, methods, and applications. *PLoS Computational Biology*, 12(6): e1004966. Online: doi:10.1371/journal.pcbi.1004966.

Löwe, S., O'Connor, P., and Veeling, B. S. (2020). Putting an End to End-to-End: Gradient-Isolated Learning of Representations. *33rd Conference on Neural Information Processing Systems (NeurIPS 2019)*, Vancouver, Canada. Online arXiv:1905.11786v3 [cs.LG].

Luczak, A., McNaughton, B. L., and Kubo, Y. (2022). Neurons learn by predicting future activity. *Nature Machine Intelligence*, 4: 62–72.

Ludwig, D. (2023). The functions of consciousness in visual processing. *Neuroscience of Consciousness*, 2023: 1–14.

Maass, W., and Schmitt, M. (1999). On the complexity of learning for spiking neurons with temporal coding. *Information and Computation*, 153: 26–46.

Maass, W., and Zador, A. (1999). Dynamic stochastic synapses as computational units. *Neural Computation*, 11: 903–917.

Madden, M. B. et al. (2022). A role for the claustrum in cognitive control. *Trends in Cognitive Science*, 26: 1133–1152. Online: doi:10.1016/j.tics.2022.09.006.

Maguire, E. A. et al. (2000). Navigation-related structural change in the hippocampi of taxi drivers. *PNAS*, 97: 4398–4403.

Marder, E., and Bucher, D. (2001). Central pattern generators and the control of rhythmic movements. *Current Biology*, 11: R986–R996.

Martin, A. R., Brown, D. A., Diamond, M. E., Cattaneo, A., and De-Miguel, F. F. (2021). *From Neuron to Brain*, 6th edn. New York, NY: Oxford University Press.

Mascetti, G. G. (2016). Unihemispheric sleep and asymmetrical sleep. *Nature and Science of Sleep*, 8: 221–238.

McIntosh, R. D., and Schenk, T. (2009). Two visual streams for perception and action. *Neuropsychologia*, 47: 1391–1396.

Mei, J., Muller, E., and Ramaswamy, S. (2022). Informing deep neural networks by multiscale principles of neuromodulatory systems. *Trends Neuroscience,* 45: 237–250. Online: doi:10.1016/j.tins.2021.12.008.

Merabet, L. B., Swisher, J. D., McMains, S. A., Halko, M. A., Amedi, A., Pascual-Leone, A., and Somers, D. C. (2006). Combined activation and deactivation of visual cortex during tactile sensory processing. *Journal of Neurophysiology*, 97: 1633–1641.

McCulloch, W. S., and Pitts, W. (1943). A logical calculus of the ideas immanent in nervous activity. *Bulletin of Mathematical Biophysics*, 5: 115–133.

McGilchrist, I. (2019). *The Master and his Emissary*. New Haven, CT: Yale University Press.

McGilchrist, I. (2022). *The Matter with Things*. London: Perspectiva Press.

McClelland, J. L., and Rumelhart, D. E. (1988). *Explorations in Parallel Distributed Processing*. Cambridge, MA: Bradford Book, MIT Press.

Meriney, S. D., and Fanselow, E. E. (2019). *Synaptic Transmission*. London: Academic Press.

Migliore, M., and Shepherd, G. (2005). An integrated approach to classifying neuronal phenotypes. *Nature Review Neuroscience*, 6: 810–818.

Mineroff, Z., Blank, I. A., Mahowald, K., and Fedorenko, E. (2018). A robust dissociation among the language, multiple demand, and default mode networks. *Neuropsychologia*, 119: 501–511.

Minh, V. et al. (2013). Playing Atari with Deep Reinforcement Learning. *NIPS Deep Learning Workshop 2013*. Online: arXiv:1312.5602v1 [cs.LG].

Minsky, M. (1988). *The Society of Mind*. New York, NY: Simon & Shuster.

Minsky, M. L., and Papert, S. A. (first edition 1969, second edition 1988). *Perceptrons*. Cambridge, MA: MIT.

Moger-Reischer, R. Z., et al. (2023). Evolution of a minimal cell. *Nature*, 620: 122–127. Online: doi:10.1038/s41586-023-06288-x.

Moini, J., and Piran, P. (2020). *Functional and Clinical Neuroanatomy*. London: Academic Press, Elsevier.

Moore, D. S. (2017). *The Developing Genome.* New York, NY: Oxford UP.

Moulin-Frier, C. et al. (2017). Embodied Artificial Intelligence through Distributed Adaptive Control. Updated version of the paper published at the ICDL-Epirob 2017 conference (Lisbon, Portugal). Online: arXiv:1704.01407 [cs.AI].

Mountcastle, V. B. (1998). *Perceptual Neuroscience.* Cambridge, MA: Harvard University Press.

Muller, L., Chavane, F. J., and Sejnowski, T. J. (2018). Cortical travelling waves: Mechanisms and computational principles. *Nature Review Neuroscience*, 19: 255–268.

Nabavi, S. (2014). Engineering a memory with LTD and LTP. *Nature*, 511: 348–352.

Nani, A., Manuello, J., Mancuso, L., Liloia, D., Costa, T., and Cauda, F. (2019). The neural correlates of consciousness and attention. *Frontier in Neuroscience*, 13: 1169. Online: doi:10.3389/fnins.2019.01169.

Natschläger, T., Maass, W., and Zador, A. (2001). Efficient temporal processing with biologically realistic dynamic synapses. *Network*, 12: 75–87.

Neal, R. M., and Hinton, G. E. (1998). A view of the EM algorithm that justifies incremental, sparse, and other variants. In M. I. Jordan (ed.), *Learning in Graphical Models*, 355–368. Dordrecht: Kluwer.

Neelakantan, A. et al. (2022). Text and Code Embeddings by Contrastive Pre-Training. Online: arXiv:2201.10005 [cs.CL]. doi:10.48550/arXiv.2201.10005.

Newell, A. (1990). *Unified Theories of Cognition.* Cambridge, MA: Harvard University Press.

Newell, A. (1992). Precis of unified theories of cognition. *Behavioral and Brain Science*, 15: 425–492.

Nurse, P. (2021). *What is Life?* London: David Fickling Books.

Ocklenburg, S., and Gunturkun, O. (2017). *The Lateralized Brain.* Cambridge, MA: Academic Press.

Olshausen, B. A., and Field, D. J. (2004). Sparse coding of sensory inputs. *Current Opinion in Neurobiology*, 14: 481–487.

Oord, van den A., Li, Y., and Vinyals, O. (2019). Representation Learning with Contrastive Predictive Coding. Online arXiv:1807.03748v2 [cs.LG].

OpenAI, GPT-4 Technical Report. (2023). Online arXiv:2303.08774v3 [cs.CL].

Pascual, O. et al. (2005). Astrocytic purinergic signaling coordinates synaptic networks. *Science*, 310: 113–116.

Pashler, H. (1994). Dual-task interference in simple tasks: Data and theory. *Psychological Bulletin*, 116: 220–244.

Pasupathy, A., and Miller, E. K. (2005). Different time courses of learning-related activity in the prefrontal cortex and striatum. *Nature*, 433: 873–876.

Parisien, C., Anderson, C. H., and Eliasmith, C. (2008). Solving the problem of negative synaptic weights in cortical models. *Neural Computation,* 20: 1473–1494.

Parrish, A. et al. (2022). BBQ: A Hand-Built Bias Benchmark for Question Answering. Online: arXiv:2110.08193v2 [cs.CL].

Paulk, A. C. et al. (2022). Large-scale neural recordings with single neuron resolution using Neuropixels probes in human cortex. *Nature Neuroscience,* 25: 252–263.

Pearl, J. (1988). *Probabilistic Reasoning in Intelligent Systems.* San Francisco, CA: Morgan Kaufman Publishers.

Pearl, J. (2009). *Causality*. Cambridge: Cambridge University Press.

Pearl, J., and Mackenzie, D. (2018). *The Book of Why*. London: Penguin Random House UK.

Peper, A. (2020). A general theory of consciousness I: Consciousness and adaptation. *Communicative & Integrative Biology*, 13: 6–21.

Pepperberg, I. M. (2006). Cognitive and communicative abilities of Grey parrots. *Applied Animal Behaviour Science*, 100: 77–86.

Perea, G., Navarrete, M., and Araque, A. (2009). Tripartite synapses. *Trends in Neurosciences*, 32: 421–431.

Pereda, A. E. (2014). Electrical synapses and their functional interactions with chemical synapses. *Nature Review Neuroscience,* 15: 250–263.

Perin, R., Berger, T. K., and Markram, H. (2011). A synaptic organizing principle for cortical neuronal groups. *PNAS*, 108: 5419–5424.

Pettito, L. A., Zatorre, R. J., Gauna, K., Nikelski, E. J., Dostie, D., et al. (2000). Speech-like cerebral activity in profoundly deaf people processing signed languages. *Proceeding of the National Academy Science of the United States America*, 97: 13961–13966.

Pfeifer, R., and Bongard, J. (2007). *How the Body Shapes the Way We Think*. Boston, MA: Bradford Books.

Piaget, J. (2023). "Intelligence is what you use when you don't know what to do: When neither innateness nor learning has prepared you for the particular situation." I have tried to find the source of this quote without success, if you know the source please contact me.

Pilly, P.K. and Grossberg, S. (2013). Spiking neurons in a hierarchical self-organizing map model can learn to develop spatial and temporal properties of entorhinal grid cells and hippocampal place cells. PLOS ONE 8: e60599: 1–21. Online: doi: 10.1371/journal.pone.0060599.

Plate, T. A. (2003). *Holographic Reduced Representation: Distributed Representation of Cognitive Structure*. Stanford, CA: CSLI.

Potier, S., Lieuvin, M., Pfaff, M., and Kelber, A. (2020). How fast can raptors see? *Journal of Experimental Biology*, 223: jeb209031. Online: doi:10.1242/jeb.209031.

Power, A. et al. (2022). Grokking, generalization beyond overfitting on small algorithmic datasets. Online: arXiv:2201.02177v1 [cs.LG].

Preston, J. (1997). *Feyerabend*. Oxford: Polity, Blackwell.

Ptito, M., Bleau, M., and Bouskila, J. (2021). The retina: A window into the brain. *Cells*, 10: 3269. Online: doi:10.3390/cells10123269.

Purdy, S. (2016). Encoding data for HTM systems. Online: arXiv: 1602.05925.

Radford, A., Narasimhan, K., Salimans, T., and Sutskever, I. (2018). Improving Language Understanding by Generative Pre-Training. *Online: OpenAI*.

Raichle, M. E. (2009). A paradigm shift in functional brain imaging. *The Journal of Neuroscience*, 29: 12729–12734.

Raichle, M. E. (2010). Two views of brain function. *Trends in Cognitive Sciences*, 14: 1364–6613.

Raichle, M. E. (2015). The brain's default mode network. *Annual Review of Neuroscience*, 38: 433–447. Online: doi:10.1146/annurev-neuro-071013-014030.

Rakic, P. (2008). Confusing cortical columns. *PNAS*, 105: 12099–12100.

Ramezanian-Panahi, M. et al. (2022). Generative models of brain dynamics. *Frontier in Artificial Intelligence*, 5: 807406. Online: doi:10.3389/frai.2022. 807406.

Real, E., Liang, C., So, D. R., and Le, Q. V. (2020). AutoML-Zero: Evolving Machine Learning Algorithms from Scratch. *Proceeding of 37th International Conference on Machine Learning*, Vienna, Austria, PMLR 119. Online: arXiv:2003.03384v2 [cs.LG].

Redinbaugh, M. J. et al. (2019). Thalamus modulates consciousness via layer-specific control of cortex. *Neuron*, 106: 66–75.

Richards, J., and Gumz, M. L. (2012). Advances in understanding the peripheral circadian clocks. *The FASEB Journal*, 26: 3602–3613.

Ringach, D. L. (2003). Neuroscience: States of mind. *Nature*, 425: 912–913.

Ritter, F. E., Tehranchi, F., and Oury, J. D. (2018). ACT-R: A cognitive architecture for modeling cognition. *Wiley Interdisciplinary Reviews: Cognitive Science*, 10: e1488.

Robertson, B. J. (2015). *Holacracy*. New York, NY: Henry Holt and Company.

Rockel, A. J., Hiorns, R. W., and Powell, T. P. S. (1980). The basic uniformity in structure of the neocortex. *Brain*, 103: 221–244.

Rogers, L. J., Zucca, P., and Vallortigara, G. (2004). Advantages of having a lateralized brain. *Proceedings of the Royal Society B*, 271: S420–S422.

Rogers, L. J., Vallortigara, G., and Andrew, R. J. (2013a). *Divided Brains*. Cambridge: Cambridge University Press.

Rogers, L. J., Rigosi, E., Frasnelli, E., and Vallortigara, G. (2013b). A right antenna for social behaviour in honeybees. *Scientific Reports*, 3: 2045. Online: doi:10.1038/srep02045.

Rogers, L. J., and Vallortigara, G., eds. (2017). *Lateralized Brain Functions*. New York, NY: Springer Science.

Rosenblatt, F. (1962). *Principles of Neurodynamics, Perceptrons and the Theory of Brain Mechanisms*. New York, NY: Spartan Books. Available as Report #1196-0-8, military report, 1961-03-15.

Rosenbloom, P. S., Laird, J. E., and Newell, A. (1993). *The Soar Papers*. Cambridge, MA: MIT Press.

Rossi, E. et al. (2022). Riding the (brain) waves! Using neural oscillations to inform bilingualism research. *Bilingualism: Language and Cognition*, 26: 202–215. Online: doi:10.1017/ S1366728922000451.

Rowland, D. C., Roudi, Y., Moser, M-B., and Moser, E. I. (2016). Ten Years of Grid Cells. *Annual Review of Neuroscience*, 39: 19–40.

Rumelhart, D. E., Hinton, G. E., and Williams, R. J. (1986a). Learning representations by back-propagating errors. *Nature*, 323: 533–536.

Rumelhart, D. E., McClelland, J., and the PDP Research Group (1986b). *Parallel Distributed Processing, Explorations in the Microstructure of Cognition, Volume 1: Foundations, Volume 2: Psychological and Biological Models*. Cambridge, MA: Bradford Book, MIT Press.

Russell, S., and Norvig. P. (2020). *Artificial Intelligence: A Modern Approach*, 4th edn. London: Pearson.

Russo, L. (2000). *The Forgotten Revolution*. Berlin: Springer Verlag.

Sacramento, J., Costa, R. P., Bengio, Y., and Senn, W. (2017). Dendritic error backpropagation in deep cortical microcircuits. Online arXiv:1801.00062v1 [q-bio.NC].

Scellier, B., and Bengio, Y. (2017). Equilibrium propagation: Bridging the gap between energy-based models and backpropagation. *Frontier in Computational Neuroscience*, 11: 24. Online: doi:10.3389/fncom.2017.00024.

Scherzer, S. et al. (2022). Ether anesthetics prevents touch-induced trigger hair calcium-electrical signals excite the Venus flytrap. *Science Report,* 12: 2851. Online: doi:10.1038/s41598-022-06915-z.

Schmidhuber, J. (2015a). Deep learning in neural networks: An overview. *Neural Networks*, 61: 85–1117.

Schmidhuber, J. (2015b). Critique of Paper by "Deep Learning Conspiracy". *Nature*, 521: 436–444. Online: https://people.idsia.ch/~juergen/deep-learning-conspiracy.html.

Schmidhuber, J. (2022). Annotated History of Modern AI and Deep Learning. Technical Report IDSIA- 22-22. Online, arXiv:2212.11279v2 [cs.NE].

Schölkopf, B. (2019). Causality for Machine Learning. Online arXiv:1911.10500v2 [cs.LG].

Schölkopf, B. et al. (2021). Towards Causal Representation Learning. Online arXiv:2102.11107v1 [cs.LG].

Schönauer, M., Geisler, T., and Gais, S. (2013). Strengthening procedural memories by reactivation in sleep. *Journal of Cognitive Neuroscience*, 26: 143–153.

Sheth, B. R., and Young. R. (2016). Two visual pathways in primates based on sampling of space: Exploitation and exploration of visual information. *Frontier in Integrative Neuroscience*, 10: 37. Online: doi:10.3389/fnint.2016.00037.

Shimoni et al. (2019). An Evaluation Toolkit to Guide Model Selection and Cohort Definition in Causal Inference. Online arXiv:1906.00442v1 [stat.ML].

Shoham, S., O'Connor, D. H., and Segev, R. (2006). How silent is the brain: Is there a "dark matter" problem in neuroscience? *Journal of Comparative Physiology A,* 192: 777–784. Online: doi:10.1007/s00359-006-0117-6.

Segal, M. (2023). Why the brain is so noisy. Nautilus online: https://getpocket.com/explore/item/why-the-brain-is-so-noisy.

Semyanov, A., and Verkhratsky, A. (2021). Astrocytic processes. *Trends in Neurosciences*, 44: 781–792.

Serrano-Gottarredona, T., Linares-Barranco, B., and Andreou, A. G. (1998). *Adaptive Resonance Theory Microchips*. New York, NY: Springer.

Seth, A. (2021). *Being You*. Boston, MA: Dutton/Penguin.

Sheheitli, H., and Jirsa, V. K. (2020). A mathematical model of ephaptic interactions in neuronal fiber pathways. *Network Neuroscience*, 4: 595–610.

Shinbrot, T., and Young, W. (2008). Why decussate? Topological constraints on 3D wiring. *The Anatomical Record*, 291: 1278–1292.

Shoham, S., O'Connor, D. H., and Segev, R. (2006). How silent is the brain. *Journal of Comparative Physiology A*, 192: 777–784.

Si, K., Choi, Y.-B., White-Grindley, E., Majumdar, A., and Kandel, E. R. (2010). Aplysia CPEB can form prion-like multimers in sensory neurons that contribute to long-term facilitation. *Cell,* 140: 421–435.

Simon, D. (2013). *Evolutionary Optimization Algorithms*. Hoboken, NJ: J Wiley & Sons.

Sinz, F. H. et al. (2019). Engineering a less artificial intelligence. *Neuron*, 103: 967–979.

Smith, G. B. et al. (2018). Distributed network interactions and their emergence in developing neocortex. *Nature Neuroscience,* 21: 1600–1608. doi:10.1038/s41593-018-0247-5.

Smith, S. D. G., Escobedo, R., Anderson, M., and Caudell, T. P. (1997). A deployed engineering design retrieval system using neural networks. *IEEE Transactions on Neural Networks*, 8: 847–851.

Song, Y., Lukasiewicz, T., Xu, Z., and Bogacz, R. (2020). Can the brain do backpropagation? *Advances Neural Information Processing System*, 33: 22566–22579.

Spalding, K. L. et al. (2013). Dynamics of hippocampal neurogenesis in adult humans. *Cell*, 153: 1219–1227.

Sporns, O. (2010). *Networks of the Brain*. Cambridge, MA: MIT Press.

Spruston, N. (2008). Pyramidal neurons. *Nature Reviews*, 9: 206–221.

Squire, L. R. (2004). Memory systems of the brain: A brief history and current perspective. *Neurobiology of Learning and Memory*, 82: 171–177.

Squire, L. R. (2009). The legacy of patient H.M. for neuroscience. *Neuron*, 61: 6–9.

Squire, L. R., and Kandel, E. R. (2009). *Memory*, 2nd edn. Greenwood Village, CO: Roberts & Company.

Srivastava, A. et al. (2022). Beyond the Imitation Game benchmark (BIG-bench). Online: arXiv:2206.04615v2 [cs.CL].

Stiefel, K. M., and Ermentrout, G. B. (2016). Neurons as oscillators. *Journal of Neurophysiology*, 116: 2950–2960. Online: doi:10.1152/jn.00525.2015.

Sterling, P., and Laughlin, S. (2015). *Principles of Neural Design*. Cambridge, MA: MIT Press.

Stiles, J., and Jernigan, T. L. (2010). The basics of brain development. *Neuropsychology Review*, 20: 327–348.

Streng, M. L., Popa, L. S., and Ebner, T. J. (2018). Complex spike wars. *The Cerebellum*, 17: 735–746.

Striedter, G. F., and Northcutt, R. G. (2020). *Brains Through Time*. New York, NY: Oxford University Press.

Stuart, G. J., and Sakmann, B. (1994). Active propagation of somatic action potentials into neocortical pyramidal cell dendrites. *Nature*, 367: 69–72.

Stuart, G. J., Spruston, N., Sakmann, B., and Häusser, M. (1997). Action potential initiation and backpropagation in neurons of the mammalian CNS. *Trends in Neuroscience*, 20: 125–131.

Stuart, G. J., and Spruston, N. (2015). Dendritic integration: 60 years of progress. *Nature Neuroscience*, 18: 1713–1721.

Sutton, R. S., and Barto, A. G. (2018). *Reinforcement Learning: An Introduction*, 2nd edn. Cambridge, MA: MIT Press. Online: https://www.andrew.cmu.edu/course/10-703/textbook/BartoSutton.pdf.

Suzgun, M. et al. (2022). Challenging BIG-Bench tasks and whether chain-of-thought can solve them. Online arXiv:2210.09261v1 [cs.CL].

Taber, K. H., and Hurley, R. A. (2014). Volume Transmission in the brain: Beyond the synapse. *Journal of Neuropsychiatry Clinical Neuroscience*, 26: 1–4. Online: doi:10.1176/appi.neuropsych.13110351.

Tan, X., and Shi S-H. (2013). Neocortical neurogenesis and neuronal migration. *Wiley Interdisciplinary Review Developmental Biology*, 2: 443–459.

Tang, J. C. Y. et al. (2023). Dynamic behaviour restructuring mediates dopamine-dependent credit assignment. *Nature*. Online: doi:10.1038/s41586-023-06941-5.

Thagard, P. (2012). Cognitive architectures. In K. Frankish and W. Ramsay (eds.), *The Cambridge Handbook of Cognitive Science*, 50–70. Cambridge: Cambridge University Press.

Thorpe, S. J. (1990). Spike arrival times: A highly efficient coding scheme for neural networks. In R. Eckmiller, G. Hartmann & G. Hauske (Eds.), *Parallel processing in neural systems and computers*, 91–94. North-Holland Elsevier: Amsterdam.

Thorpe, S., and Gautrais, J. (1996). Rapid Visual Processing using Spike Asynchrony. *NeurIPS Proceedings: Advances in Neural Information Processing Systems*, 9.

Thorpe, S., Delorme, A., and Van Rullen, R. (2001). Spike-based strategies for rapid processing. *Neural Network*, 14: 715–725. Online: doi:10.1016/s0893-6080(01)00083-1.

Thorpe, S. (2023). Timing, Spikes and the Brain. In R. Lestienne and P. A. Harris (eds.), *Time and Science*, 2: 208–236. Singapore: World Scientific.

Tononi, G. (2004). An information integration theory of consciousness. *BMC Neuroscience*, 5: 42. Online: doi:10.1186/1471-2202-5-42.

Tononi, G., and Koch, C. (2008). The neural correlates of consciousness: An update. *Annals of the New York Academy of Science*, 1124: 239–261.

Tse, D. et al. (2007). Schemas and Memory Consolidation. *Science*, 316: 76–82.

Umbach, G. et al. (2020). Time cells in the human hippocampus and entorhinal cortex support episodic memory. *PNAS*, 117: 28463–28474.

Van Essen, D. C. (2020). A 2020 view of tension-based cortical morphogenesis. *PNAS*, 117: 32868–32879.

Vaswani, A., Shazeer, N., Parmar, N., Uszkoreit, J., Jones, L., Gomez, A. N., Kaiser, L., and Polosukhin, I. (2017). Attention Is All You Need. *31st Conference on Neural Information Processing Systems (NIPS 2017)*.

Wan, W., Kubendran, R., Schaefer, C. et al. (2022). A compute-in-memory chip based on resistive random-access memory. *Nature*, 608: 504–512. Online: doi:10.1038/s41586-022-04992-8.

Watts, D. J., and Strogatz, S. H. (1998). Collective dynamics of 'small-world' networks. *Nature*, 393: 440–442.

Weber, S. N., and Sprekeler, H. (2018). Learning place cells, grid cells and invariances with excitatory and inhibitory plasticity. *eLife*, 7: e34560, online: doi:10.7554/eLife.34560.

Wei, J. (2022). 137 emergent abilities of large language models. Online: www.jasonwei.net/blog/emergence.

Wei, J. et al. (2022). Emergent Abilities of Large Language Models. *Transactions on Machine Learning Research (08/2022)*. Online: https://openreview.net/forum?id=yzkSU5zdwD.

Wei, J. et al. (2023). Chain-of-Thought Prompting Elicits Reasoning in Large Language Models. 36th NeurIPS. Online: arXiv:2201.11903v6 [cs.CL].

Weiss, S. A., and Faber, D. S. (2010). Field effects in the CNS play functional roles. *Frontiers in Neural Circuits*, 4. Online: doi:10.3389/fncir.2010.00015.

Weng. J. et al. (2001). Autonomous mental development by robots and animals. *Science* 291: 599–600. Online: doi:10.1126/science.291.5504.599.

Werbos, P. (1974). Beyond Regression: New Tools for Prediction and Analysis in the Behavioral Science. Thesis (Ph. D.). Applied Mathematics. Harvard University.

Whittington, J. C. R., and Bogacz, R. (2019). Theories of error back-propagation in the brain. *Trends in Cognitive Sciences*, 23: 235–250. Online: doi:10.1016/j.tics.2018.12.005.

Wolf, S. (2021). Recurrent dynamics in the cerebral cortex. *PNAS*, 118: e2101043118. Online: doi:10.1073/pnas.2101043118.

Yin, W. et al. (2020). A petascale automated imaging pipeline for mapping neuronal circuits with high- throughput transmission electron microscopy. *Nature Communications*, 11: 4949. Online: doi:10.1038/s41467-020-18659-3.

Yu, A. J., and Dayan, P. (2005). Uncertainty, neuromodulation, and attention. *Neuron*, 46: 681–692. Online: doi:10.1016/j.neuron.2005.04.026.

Yuen, S., Ezard, T. H. G., and Sobey, A. J. (2023). Epigenetic opportunities for evolutionary computation. *Royal Society of Open Science*, 10: 221256. Online: doi:10.1098/rsos.221256.

Yuste, R. (2015). From the neuron doctrine to neural networks. *Nature Review Neuroscience*, 16: 487–497.

Zador, A. M., and Dobrunz, L. E. (1997). Dynamic Synapses in the Cortex. *Neuron*, 19: 1–4.

Zador, A. M. (2019). A critique of pure learning and what artificial neural networks can learn from animal brains. *Nature Communication*, 10: 3770. Online: doi:10.1038/s41467-019-11786-6.

Zador, A. et al. (2023). Catalyzing next-generation Artificial Intelligence through NeuroAI. *Nature Communications*, 14: 1597. Online: doi:10.1038/s41467-023-37180-x.

Ziemke, T. (2003). What's that thing called embodiment? *Proceedings of the Annual Meeting of the Cognitive Science Society*, 25: 1305–1310.

Zoubine, M. N., Ma, J. Y., Smirnova, I. V., Citron, B. A., and Festoff, B. W. (1996). A molecular mechanism for synapse elimination. *Developmental Biology*, 179: 447–457.

Zuckerman, G. (2019). *The Man Who Solved the Market*. New York, NY: Penguin Books.

Zyarah, A. M., and Kudithipudi, D. (2019). Neuromorphic architecture for the hierarchical temporal memory. *IEEE Transaction on Emerging Topics in Computational Intelligence*, 3: 4–14.

Zyarah, A. M., Gomez, K., and Kudithipudi, D. (2020). Neuromorphic system for spatial and temporal information processing. *IEEE Transaction on Computers*, 69: 1099–1112.

Author: Eitan Michael Azoff

My education (BEng from UCL in electronic and electrical engineering, MSc from the University of London in solid state physics, and PhD from the University of Sheffield in solid state electronics) led to research work in novel semiconductor device simulation for Ministry of Defense contracts at the University of Sheffield, and then Rutherford Appleton Laboratory on UK and EU research contracts.

In 1989, I joined a startup in the University of Nottingham Science Park, which included working with neural networks, funded by an award from the UK government Department of Trade and Industry. On Black Wednesday when the UK left the ERM our startup's funder backed out and I started working as a consultant through my company Netnumerics. With Perot Systems (since acquired by Dell) I developed electricity generation price forecasts for East Midlands Electricity (since acquired by Powergen) using neural networks. I also built a Microsoft Excel add-in, Prognostica, for time series forecasting. During this time, I engaged with partners, including a boutique investment house, to apply neural networks to time series forecasting of financial markets. My models generated 55% accuracy in predicting the direction of major financial indices like the FTSE 100. We considered this not good enough (kind of mistakenly – interestingly I recently read that Jim Simons at Renaissance had similar statistics and with deep enough pockets could monetize it). I subsequently wrote up my knowledge of neural networks in a book (Azoff, 1994), out of print but due to re-appear.

In the past two decades I've worked as a high-tech industry analyst, mostly at Informa business Ovum, and after an independent break currently back with Informa business Omdia as chief analyst in the cloud and data center practice. This book arose in my time as an independent consultant.

My interest in AI followed the invention of backpropagation, the breakthrough and excitement it stirred touched me around 1988. I joined a newly formed UK group with membership drawn from industry and academia, the Neural Computing Applications Forum, which led to the formation of a research journal published by Springer: *Neural Computing and Applications*, serving on the editorial board. After Netnumerics in subsequent decades I kept my interest in AI on hold but began to cover it as an industry analyst with the emergence of deep learning in 2010–2012, the rise of AI accelerator processors, and more recently, the breakthroughs with generative AI and LLM.

Index

AI
  ACT-R 69
  ACT-R as subsymbolic model 70
  adaptive resonance theory (ART)
      5, 72
  AI hardware 105
  AI level setting 1
  AI machine enhanced with novel
      sensors and limbs 144
  AI winter 1
  aligning neuroscience and AI
      research 119
  animal-level AI (ALAI) 118, 122
  animal or human level AI (A/HLAI)
      119
  artificial general intelligence (AGI)
      1, 137
  artificial versus living neuron 131
  as statistical machine driven by
      external data 100
  backpropagation 1, 5, 7
  backpropagation in the brain 93
  Bayesian probabilistic learning 3
  beyond HLAI 144
  biologically plausible models 93
  biologically plausible neural
      networks 98
  biological neural network
      (BNN) 111
  brain learning (BL) algorithm 93
  building HLAI 130
  chatbot to chatbot communication 127
  ChatGPT 85
  cognitive architecture 65, 134
  cognitive-emotional-motor
      (CogEM) 75
  cognitive model theory 66
  curse of dimensionality 9

  damped harmonic oscillation
      (DHO) 76
  deep learning 2, 8, 73
  deep learning limitations 73, 120
  deep learning neural networks
      (DLNN) 5, 83
  DLNN model of a single human
      neuron 99
  DLNN replicates a neuron 130
  emergent property definition 87
  engineered brain architecture 65
  engineered HLAI (engHLAI) 117
  equilibrium propagation (EP) 94
  evolving AI systems 63
  evolving intelligent systems 136
  expectation-maximization (EM)
      algorithm 99
  feedforward neural network 6
  few-shot learning 92
  general-purpose graphics processing
      unit (GPGPU) 8
  generative adversarial net (GAN) 84
  generative AI 2, 5, 83
  generative pre-trained transformer
      (GPT) 85
  harmonic oscillator recurrent neural
      networks 76
  harmonic oscillator RNN
      (HORN) 76
  hierarchical temporal memory
      (HTM) 79
  HLAI as computer simulation 121
  holographic reduced representation
      103
  human-level AI (HLAI) 1
      levels of increasingly intelligent
          systems - version 1 122
      three approaches 117

human-like human-level AI
    [(HL)²AI] 117
hybridHLAI (H²LAI) 118
hybrid systems 3
hyperdimensional computing
    (HDC) 103, 135
inference 9, 107
key attributes and tests of an A/
    HLAI system 137, 138
key attributes of an HLAI
    system 140
language in HLAI research 122
large language model (LLM)
    85, 135
levels of increasingly intelligent AI
    systems - version 2 144
LLM chain-of-thought reasoning 89
LLM emergent properties 86
LLM fine-tuning 85
LLM hallucinations 92
LLM retrieval augmented
    generation (RAG) 85
LLM solves high-level
    mathematics 92
machine learning 2
memory in HLAI systems 127
narrow AI 2
natural language processing 3, 59
natural selection (evolutionary)
    algorithms 96
networks of coupled oscillators 78
networks of neuron ensembles 135
neural ensembles as holocracy
    circles 129
neural networks history 1, 5
neuroAI grand challenge
    the embodied Turing test 120
neuromodulation 135
neuromodulation in AI 131
neurorobotics (embodied AI) 62
neuro-symbolic 59
parallel distributed processing
    (PDP) 6
PDCA cycle 134
perceptron 6
preset directives 119
recurrent neural network (RNN) 76
reinforcement learning and
    neuroscience 96
reinforcement learning (RL) 95
self-supervised algorithm 89
semantic pointer architecture 71
simulation of an HLAI system 130

singularity 2
small world network 63
Soar 67
sparse distributed memory 79, 103
sparse distributed representation
    (SDR) 79
sparse interconnections and sparse
    activity of neurons 83
spike-based computation 101
spike timing-dependent plasticity
    (STDP) 107
spiking neural networks 105
spiking neuron (in ART) 75
stability-plasticity dilemma 72
symbolic and expert systems 3
symbolist 66
testing an HLAI system 140
thousand brains theory 79
training 5, 106
transcendent AI 2, 118, 144
transformer 84
Turing test 120
unsupervised algorithm (in ART) 74
vector embedding 92
verbal thinking versus visual
    thinking 123
visual processing in HLAI 135
visual thinking as the earliest form
    of intelligence 122
zero-shot learning 92
Amazon Web Services (AWS) 84
AMD Xilinx 83

Baldwin effect 96
biology
    epigenetics 33, 36, 97
    extended evolutionary synthesis 97
    non-genetic inheritance 97
    protein as packaged algorithm 131
    the modern synthesis 97
brain
    2D sensory input to 3D vision 34
    2D sensory input to 3D vision 34
    acetylcholine 29
    activity, sparsity, and
        normalization 39
    akinetopsia (motion blindness) 54
    astrocyte (glial cell) 21
    autonomous systems 128
    brain folds (gyrification) 35
    brain lateralization communication
        126
    brain lateralization cross-wiring 34

brain's 3D visual processing of 2D sensory inputs 126
Brownian motion 27
chatty brain 127
chemical diffusion 27
chemical, electrical, and field neuron junctions 138
circadian clock 55
compared with digital microelectronic circuit 39
computational neuroscience 59
connecting brain lateralization, visual processing, and intelligence 126
corpus callosum 33, 35
cyclic adenosine monophosphate (cAMP) 32
Darwinian neurodynamics 97
Darwinian selection process (thoughts) 15
decussation 34
default mode network (DMN) 15, 39
diffuse decision making 128
dorsal and ventral processing 34
early brain development 36
echoic memory 31
EEG measurements 15, 19, 41
electrical communication 23
free-energy principle 99, 112
grey matter 22
haptic memory 31
hippocampus 32, 49, 56, 99
holocracy
    self-organized diffuse decision making 128
homunculus 14, 97
hormone 28
human memory 29
iconic memory 31
inside out 45, 134
internal model of the external environment 54
language came after vision 52
lateralization 33
lateralization as foundation for intelligence 126
long-term memory 30
mammalian lineages 36
mass of brain 20
master biological clock 55
memory plasticity 127
neocortex 16, 35
neural Darwinism 96

neuroanatomy 27
neuromodulation in AI 98
neuromodulator 25, 27, 42, 131
    acetylcholine, adrenaline (epinephrine), dopamine, GABA, glutamate, histamine, noradrenaline, serotonin 28
neuropeptide 28
neurotransmitter 27
new qualia 15
noradrenergic locus coeruleus 28
number of neurons in animal brains 131
outside in 45
plasticity 38
prediction machine 128
prion (CPEB) 31
processing delay 57
properties 12
Purkinje cell 26
reciprocity of neural connections 48
rules of visual processing 53
sensory memory 30
sensory thinking 51
short-term memory 30
society of mind 60, 139
sparse representations 19
spontaneous brain activity 15
spontaneous excitation in the brain 46
suprachiasmatic nuclei (SCN) 55
tension-based morphogenesis 36
time and space in the brain 55
triune brain hypothesis 11
verbal thinking 51
vision
    color is a construction 54
    perception and visual intelligence 52
    phenomenal and relational 53
visualizer object or pictorial 51
visualizer spatial 51
visual thinking 51
visual thinking versus verbal thinking 123
white matter 22
BrainChip Akida 102

causal inference 100, 134
causal inference libraries 101
cell
    eukaryotic cell 12
    gene 13

glial 20, 37, 38, 111
hapto-electric signaling 51
neuron doctrine 13
progenitor cell (stem cell) 36
prokaryotic cell 13
protein logic 13
Cerebras 84
cognitive processes 45
consciousness
    anesthesia 50, 51
    animal 49, 139
    birds 50
    blindsight 47
    Cambridge Declaration on
        Consciousness 49
    definition 47
    global workspace theory (GWT) 61
    human 46
    insects 50
    integrated information theory
        (IIT) 61
    in the flow consciousness (ITFC)
        121
    levels of 121
    mind as continuous spectrum of
        states 60
    neural correlates of consciousness
        (NCC) 47, 60
    panpsychism 61
    re-entry 48
    self-aware 121
    theory of mind 50
    thinking as moving concepts in
        thought space 124
    unconscious states 48
    understanding consciousness as
        futile endeavor 125
    unified theory of cognition 60
    visual thinking 122
Cooper, Leon
    precepts 115, 119, 125
credit assignment problem 28

DishBrain
    in vitro neurons learn to play
        Pong 111
distance metrics 104
DNA 36

Edward de Bono 97
Elon Musk 84
epigenetic 36

fetus 37
field programmable gate array (FPGA)
    83
fish operated vehicle 16
flying metaphor
    airplane versus bird 65, 119
four-tree problem 97
fruit fly Drosophilia melanogaster 55

general-purpose graphics processing
    unit (GPGPU) 83, 110
genes 36
GitHub 84
Google 84
Graphcore 84

intelligence
    emotional 54
    rational 54
    visual 54
Intel's neuromorphic computing
    research 108

Jim Simon's Renaissance
    Technologies 8

Kim Peek
    35
Kirchhoff's laws 109

memory palace 51, 82
Meta 84
Microsoft 84
MNIST pattern recognition
    benchmark 76

neuromorphic computing 105
neuromorphic processors 105
neuron 12
    action potential 12, 16, 17, 23
    action potential backpropagation 17
    axon initial segment 17
    axons 14
    chemical synapse 16
    Dale's principle 14
    dark neurons (neuronal silence) 40
    dendrite 38
    dendrite growth cones 37
    dendrites 19, 26, 82, 94, 138
    dMEC 22
    electrical spike 12
    electrical synapse 16, 24

ephaptic coupling 26
ephaptic interaction 25
excitatory and inhibitory 138
frequency of action potential 25
gap junction 16
grid 56
grid neuron 22
Hebb rule 24
inferior olive 26
ion channel 17
ion channels 17
mirror neuron 23
mixed selectivity 40
motor neuron 32
myelin 12, 22, 37
neural plasticity 14
neurotransmitter 16
neurotrophins 13
number in human brain 13
number of synapses 13
pyramidal neuron 19
REM atonia 14
sensory neuron 32
silent neurons 48
sodium-potassium pump 17
spike-timing dependent plasticity
        (STDP) 29
synapse dynamics 25
synaptotoxins 13

types 12
unmyelinated axon section 17
voltage-gated channels 17
neuron
    backpropagating action potential 20
neuroscience 11
    neuroscience data analysis methods
            applied to a understand a
            microprocessor 120
    office building analogy to
            understanding the brain 120
NeuRRAM analog chip 110
Nvidia 84, 108, 110

OpenAI 84
optogenetics 29

peripheral clocks 56
plan-do-check-act (PDCA) 124
proteins 36

Rain Neuromorphics 108
Raven's progressive matrices 104

scientific method 116, 125
sleep 56
small world network 39

trolley problem in philosophy 142